TWENTY FIRST CENTURY
science

Project Directors

Angela Hall
Emma Palmer
Robin Millar
Mary Whitehouse

Author

Philippa Gardom Hulme

THE UNIVERSITY *of York*

OXFORD
UNIVERSITY PRESS

Great Clarendon Street, Oxford OX2 6DP

Oxford University Press is a department of the University of Oxford.
It furthers the University's objective of excellence in research,
scholarship, and education by publishing worldwide in

Oxford New York

Auckland Cape Town Dar es Salaam Hong Kong Karachi
Kuala Lumpur Madrid Melbourne Mexico City Nairobi
New Delhi Shanghai Taipei Toronto

With offices in
Argentina Austria Brazil Chile Czech Republic France Greece
Guatemala Hungary Italy Japan Poland Portugal Singapore
South Korea Switzerland Thailand Turkey Ukraine Vietnam

Oxford is a registered trade mark of Oxford University Press
in the UK and in certain other countries.

First published 2012.

British Library Cataloguing in Publication Data.

Data available.

ISBN 978-0-19-913845-6

10 9 8 7 6 5 4 3 2 1

Printed in Great Britain by Bell and Bain Ltd, Glasgow.

Paper used in the production of this book is a natural, recyclable product
made from wood grown in sustainable forests. The manufacturing process
conforms to the environmental regulations of the country of origin.

Acknowledgements
Illustrations by IFA Design, Plymouth, UK, Clive Goodyer, and Q2A Media.

Author acknowledgements
Many thanks to Catherine and Sarah for checking the puzzles, and to
Barney for his inspirational ideas. Thanks to Ruth for her careful editing,
and to Les, Sophie, and Barry at OUP for all their help and patience.

Contents

About this book

Welcome to the Twenty-First Century Physics Revision Guide! This book will help you prepare for all your GCSE Physics module tests. There is one section for each of the physics modules P1–P7, as well as six sections covering Ideas about science. Each section includes several types of pages to help you revise.

Workout: These are to help you find out what you can remember already, and get you thinking about the topic. They include puzzles, flow charts, and lots of other types of questions. Work through these on your own or with a friend, and write your answers in the book. If you get stuck, look in the Factbank. The index will help you find what you need. Check your answers in the back of the book.

Factbank: The Factbanks summarise information from the module in just a few pages. For P1–P7, the Factbanks are divided into short sections, each linked to different statements in the Specification. The Ideas about science Factbanks are different. They are conversations, covering the ideas you will need to apply in different contexts. Read them aloud with a friend if you want to.

Quickfire: Sections P1–P7 each have Quickfire questions. These are short questions that cover most of the content of the module. For some questions, there is space to answer in the book. For others, you will need to use paper or an exercise book.

GCSE-style questions: These are like the questions in the module tests. You could work through them using the Factbank to check things as you go, or do them under test conditions. The answers are in the back of the book. Most sections include one 6-mark question, designed to test your ability to organise ideas, and write in clear and correct English. Use these to help you practise for this type of question in the module tests.

H In every section, content covered at Higher-tier only is shown like this.

Other help: This page and the next one include vital revision tips and hints to help you work out what questions are telling you to do. Don't skip these!

Making the most of revision

Remember, remember: You probably won't remember much if you just read this book. Here are some suggestions to help you revise effectively.

Plan your time: Work out how many days there are before your test. Then make a timetable so you know which topics to revise when. Include some time off.

Revise actively, don't just read the Factbanks. Highlight key points, scribble extra details in the margin or on Post-it notes, and make up ways to help you remember things. The messier the Factbanks are by the time you take your tests, the better!

Mind maps: try making mind maps to summarise the information in a Factbank. Start with an important idea in the middle. Use arrows to link this to key facts, examples, and other science ideas.

Test yourself on key facts and ideas. Use the Quickfire sections in this book, or get a friend to ask you questions. You could make revision cards, too. Write a question on one side, and the answer on the other. Then test yourself.

Try making up songs or rhymes to help you remember things. You could make up **mnemonics**, too, like this one for the colours in the visible part of the electromagnetic spectrum:

Richard **O**f **Y**ork **G**ave **b**attle **i**n **V**ain

Apply your knowledge: Don't forget you will need to apply knowledge to different contexts, and evaluate data and opinions. The GCSE-style questions in this book give lots of opportunities to practise these skills. Your teacher may give you past test papers, too.

Ideas about science: should not be ignored. These are vital. In your module tests, there could be questions on any of the Ideas about science you have covered so far, set in the context of most of the topics you have covered.

Take short breaks: take plenty of breaks during revision – about 10 minutes an hour works for most people. It's best not to sit still and relax in your breaks – go for a walk, or do some sport. You'll be surprised at what you can remember when you come back, and at how much fresher your brain feels!

Answering exam questions

Read the question carefully, and find the command word. Then look carefully at the information in the question, and at any data. How will they help you answer the question? Use the number of answer lines and the number of marks to help you work out how much detail the examiner wants.

Then write your answer. Make it easy for the examiner to read and understand. If a number needs units, don't forget to include them.

Six-mark questions

Follow the steps below to gain the full six marks:

- Work out exactly what the question is asking.
- Jot down key words to help your answer.
- Organise the key words. You might need to group them into advantages and disadvantages, or sequence them to describe a series of steps.
- Write your answer. Use the organised key words to help.
- Check and correct your spelling, punctuation, and grammar.

Below are examiner's comments on two answers to the question: ***"In 2011, an earthquake damaged a Japanese nuclear power station. Radioactive materials escaped into the environment. After this, the government of one European country made the decision to build no more nuclear power stations and to generate more electricity from a mix of renewable resources. Outline the arguments for and against this decision."***

The quality of written communication will be assessed.

Command words

Calculate Work out a number. Use your calculator if you like. You may need to use an equation.
Compare Write about the ways in which two things are the same, and how they are different.
Describe Write a detailed answer that covers what happens, when it happens, and where it happens. Your answer must include facts, or characteristics.
Discuss Write about the issues, giving arguments for and against something, or showing the difference between ideas, opinions, and facts.
Estimate Suggest a rough value, without doing a complete calculation. Use your science knowledge to suggest a sensible answer.
Explain Write a detailed answer that says how and why things happen. Give mechanisms and reasons.
Evaluate You will be given some facts, data, or an article. Write about these, and give your own conclusion or opinion on them.
Justify Give some evidence or an explanation to tell the examiner why you gave an answer.
Outline Give only the key facts, or the steps of a process in the correct order.
Predict Look at the data and suggest a sensible value or outcome. Use trends in the data and your science knowledge to help you.
Show Write down the details, steps, or calculations to show how to get an answer.
Suggest Apply something you have learnt to a new context, or to come up with a reasonable answer.
Write down Give a short answer. There is no need for an argument to support your answer.

Answer	Examiners' comments
They are right becoz nuclear power is v dangerus and reknewable energy is things like wind power that aren't dangerous, except the niose is terrible and they looks bad!	**Grade G** answer: this answer makes some correct points. However, the points are not well organised and it is not clear which arguments are for and which against. There are mistakes of spelling, grammar, and punctuation.
As the Japanese example shows, there are hazards linked to nuclear power stations. Even though the risk of radioactive materials escaping is small, the consequences can be devastating. This is an argument for the decision. Also, building nuclear power stations makes lots of CO_2*, a greenhouse gas.* *Another argument for the decision is that renewable energy resources are sustainable. They make little pollution. One renewable source alone may be unreliable, but with a mix they can generate electricity all the time.* *The arguments against the decision are that nuclear power is a constant, reliable source of electricity. Also, some people are against wind power because the turbines can kill birds. We don't know if renewables can supply enough electricity to meet demand.*	**Grade A/A*** answer: this answer is typical of an A/A* candidate. The arguments are made clearly and are organised logically. The candidate has referred to risk and sustainability. The spelling, punctuation, and grammar are faultless.

Equations and units

Equations

You might need to use these equations in the exam. They will be on the exam paper, so you do not need to learn them off by heart.

P1 The Earth in the Universe

distance travelled by a wave = wave speed × time

wave speed = frequency × wavelength

P3 Sustainable energy

energy transferred = power × time

power = voltage × current

$$\text{efficiency} = \frac{\text{energy usefully transferred}}{\text{total energy supplied}} \times 100\%$$

P4 Explaining motion

$$\textbf{Speed} = \frac{\text{distance}}{\text{time}}$$

$$\textbf{Acceleration} = \frac{\text{change in velocity}}{\text{time taken}}$$

Momentum = mass × velocity

Change of momentum
= resultant force × time for which it acts

Work done by a force
= force × distance moved in direction of force

Energy transferred = work done

Change in gravitational potential energy
= weight × height

Kinetic energy = ½ × mass × velocity^2

P5 Electric circuits

Power = voltage × current

$$\textbf{Resistance} = \frac{\text{voltage}}{\text{current}}$$

H

$$\frac{\text{Voltage across primary coil}}{\text{voltage across secondary coil}} = \frac{\text{number of turns in primary coil}}{\text{number of turns in secondary coil}}$$

P6 Radioactive materials

$E = mc^2$
(E = energy, m = mass lost, c = speed of light in a vacuum)

P7 Further physics

$$\textbf{Power of a lens} = \frac{1}{\text{focal length}}$$

$$\textbf{Magnification of a telescope} = \frac{\text{focal length of objective lens}}{\text{focal length of eyepiece lens}}$$

Hubble equation:
speed of recession = Hubble constant × distance

Einstein's equation:
$E = mc^2$
(E is the energy created, m is mass lost, c is the speed of light in a vacuum.)

For a fixed mass of gas:

At constant temperature:
pressure × volume = constant

At constant volume:

$$\frac{\text{pressure}}{\text{temperature}} = \text{constant}$$

At constant pressure:

$$\frac{\text{volume}}{\text{temperature}} = \text{constant}$$

Units

Length: metres (m), kilometres (km), centimetres (cm), millimetres (mm), micrometres (μm), nanometres (nm)

Mass: kilograms (kg), grams (g), milligrams (mg)

Time: seconds (s), milliseconds (ms)

Temperature: degrees Celsius (°C), Kelvin (K)

Area: cm^2, m^2

Volume: cm^3, m^3

Speed and **velocity:** m/s, km/s, km/h

Energy: joules (J), kilojoules (kJ), megajoules (MJ), kilowatt-hours (kWh), megawatt-hours (MWh)

Electric current: amperes (A), milliamperes (mA)

Potential difference/voltage: volts (V)

Resistance: ohms (Ω)

Power: watts (W), kilowatts (kW), megawatt (MW)

Radiation dose: sieverts (Sv)

Distance (astronomy): parsecs (pc)

Power of a lens: dioptres (D)

P1 The Earth in the Universe workout

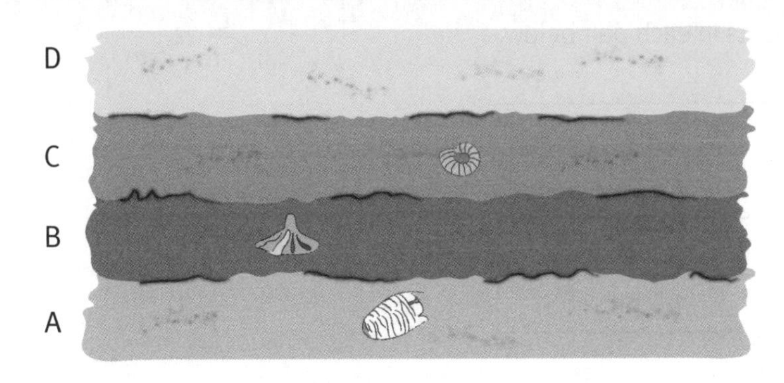

This diagram shows sedimentary rock containing fossils. Assume that this rock has never been folded.

1 Give the letter of

a the layer that contains the youngest fossils ______

b the layer made of the oldest rocks ______

c the layer made of the youngest rocks ______

d the layer in which sediments were first deposited ______

2 Write the correct numbers in the gaps. Use the numbers in the box.

12 700 10 14 thousand million
five thousand million four thousand million

a The diameter of the Earth is about ______ km.

b The Earth's oldest rocks are ______ years old.

c Seafloors spread by about ______ cm each year.

d The Sun was formed about ______ years ago.

e The 'big bang' happened about ______ years ago.

3 Write the letters of the distances below in order, smallest first. The first one has been done for you.

G								

A the diameter of the Sun

B the diameter of the Milky Way

C the distance from the Milky Way to the nearest galaxy

D the distance from the Sun to the nearest star

E the diameter of the Earth

F the diameter of Earth's Moon

G the diameter of a comet

H the diameter of the Earth's orbit

I the diameter of the Solar System

Exam tip

Remember this order of distances. You may be asked about it in the exam.

4 List one or more differences between the objects in each pair below.

a comet and asteroid ____________________

____

b moon and planet ____________________

____

c star and galaxy ____________________

____

5 Draw lines to match each type of wave to:

- the direction of vibrations
- one or more examples.

Direction of vibrations	Type of wave	Examples
vibrations are at right angles to the direction in which the wave is moving	longitudinal	P-waves
		sound waves
vibrations are in the same direction as the moving wave	transverse	S-waves
		water waves

6 Draw and label two arrows on the diagram to show the wave's **wavelength** and **amplitude**.

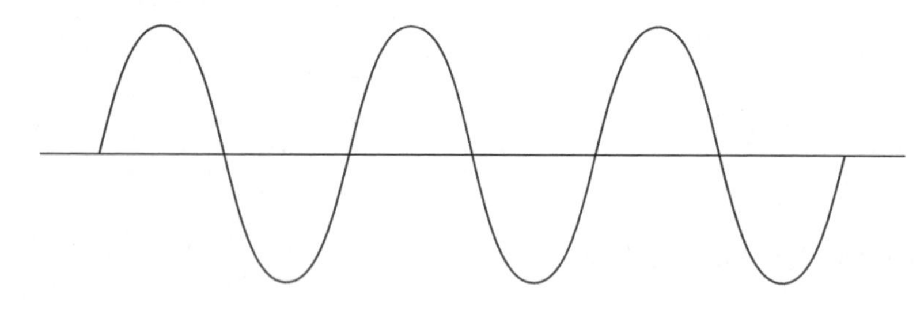

7 Calculate the speed of these waves. Include units in your answers.

a a sound wave from a firework that travels 100 m in 0.3 seconds

Answer = __________

b a seismic wave that has a wavelength of 20 km and a frequency of 0.5 Hz

Answer = __________

Exam tip

There are two equations that include wave speed. Make sure you choose the right one!

P1.1.1–3 What is the Solar System?

Our **Solar System** was formed about 5000 million years ago from clouds of gases and dust in space. It contains:

- the **Sun**, the star at the centre of the Solar System
- eight **planets**, including **Earth**
- **dwarf planets**, including Pluto
- **moons** – natural satellites that orbit some planets
- **comets** – big lumps of ice and dust
- **asteroids** – lumps of rock that are smaller than planets.

Type of body	Diameter (km)	Orbit
Planet	Mercury, 4880 (smallest) Saturn, 120 000 (biggest) Earth, 12 700	almost circular path around the Sun
Dwarf planet	Pluto, 2306	elliptical path around the Sun
Moon	Earth's moon, 3500 A moon is smaller than the planet it orbits.	circular path around a planet
Comets	a few kilometres	rush past the Sun then return to the outer Solar System
Asteroids	up to 1000; most are much smaller	almost circular, mostly between Mars and Jupiter

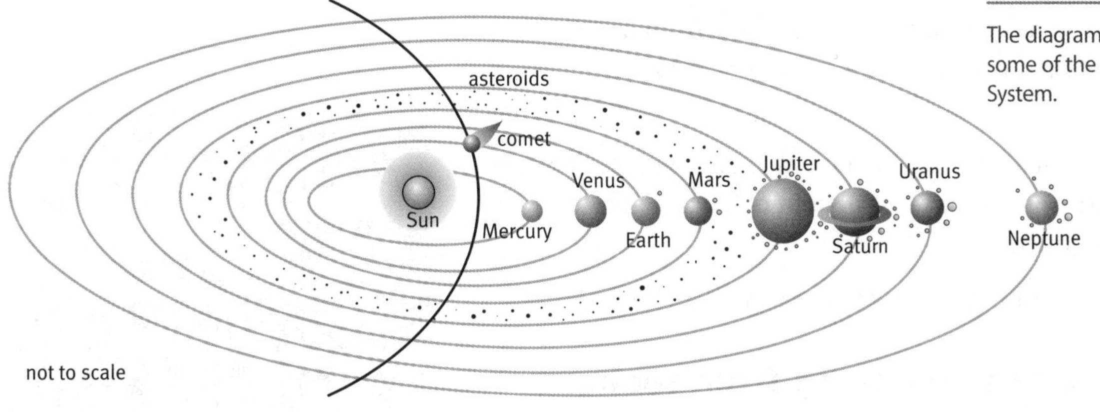

The diagram shows the orbits of some of the objects of our Solar System.

P1.1.4–5, P1.1.15–16 What is the Universe?

The Sun is a star. It is a ball of hot gases, mainly hydrogen. In the Sun, hydrogen atom nuclei join together by **nuclear fusion**. This releases energy and creates helium. All the other elements were made by fusion reactions in stars.

There are thousands of millions of stars in our **galaxy**, the Milky Way. The **Universe** is made up of thousands of millions of galaxies.

P1.1.7–14 How do we find out about stars?

Astronomers can learn about other stars and galaxies only by studying the radiation they emit. They measure the distance to stars from their **relative brightness**, or by **parallax**.

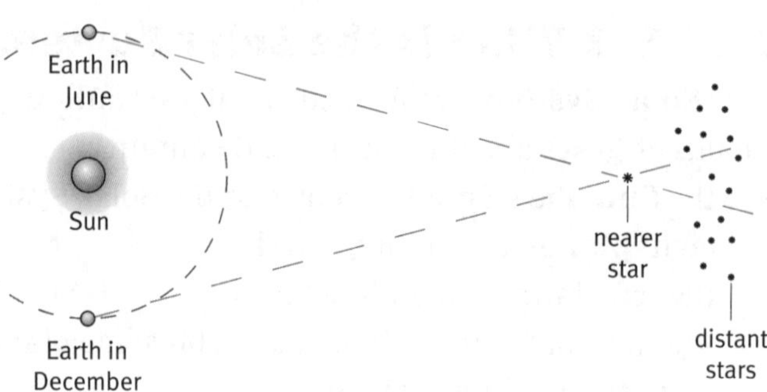

The Earth moves from one side of the Sun to the other. Nearby stars seem to move compared with the background of distant stars. The nearer a star is to Earth, the more it seems to move. This is **parallax**.

There are **uncertainties** about distances in space. This is because making accurate observations is difficult. Also, astronomers make **assumptions** when interpreting observations.

Light travels through space (a vacuum) at 300 000 km/s. So when astronomers observe a distant object they see what it looked like when light left the object – not what it looks like now. Astronomers measure distances in light-years. One light-year is the distance light travels through a vacuum in one year.

Light pollution near cities makes it difficult to observe stars. So astronomers set up telescopes away from big cities.

P1.1.17–22 How is the Universe changing?

Other galaxies are moving away from us. This is evidence that the Universe started with a 'big bang', about 14 000 million years ago.

Evidence that galaxies are moving apart comes from **redshift.** This means that light from galaxies is shifted towards the red end of the spectrum. Galaxies that are further away from us move faster than those that are closer. This is evidence that space is expanding.

It is difficult to predict how the Universe will end. This is because it is difficult to study the motion of distant objects, and to measure huge distances.

H It is difficult to work out the mass of the Universe.

P1.2.1–4 What can we learn from rocks?

Evidence from rocks tells us about the structure of the Earth. For example, the Earth must be older than its oldest rocks, which are about 4000 million years old. Geologists use **radioactive dating** to estimate rocks' ages.

Rock processes we see today explain past changes. For example, mountains are being made all the time – if not, **erosion** would wear down continents to sea level. **Sedimentation** explains why older rocks are usually found under younger rocks.

Older rocks are usually found under younger rocks. Different creatures lived at different times in the past. Their fossils can help geologists decide when rocks were formed.

P1.2.5–7 What is Wegener's theory?

In 1912, Wegener suggested his theory of **continental drift**. He explained that today's continents were once joined together. For millions of years, they have been slowly moving apart. This explains why there are mountain chains at the edges of continents.

For more details about Wegener's theory, see page 132.

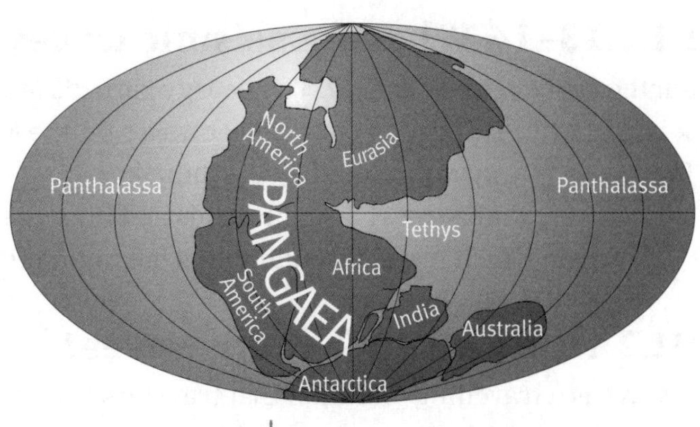

The Earth as it was 250 million years ago.

P1.2.8–10 What is seafloor spreading?

Scientists detected **ocean ridges**, which are lines of mountains under the sea. New ocean floor is made at these ridges. So seafloors spread by about 10 cm a year.

New ocean floor is made when material from the solid mantle rise slowly. Some of the mantle material melts to form magma. Movements in the mantle, caused by convection, pull an ocean ridge apart. Magma erupts and cools to make new rock.

The new rock has a symmetrical pattern of magnetic stripes. This is because as magma solidifies, it becomes magnetised in the direction of the Earth's magnetic field at the time.

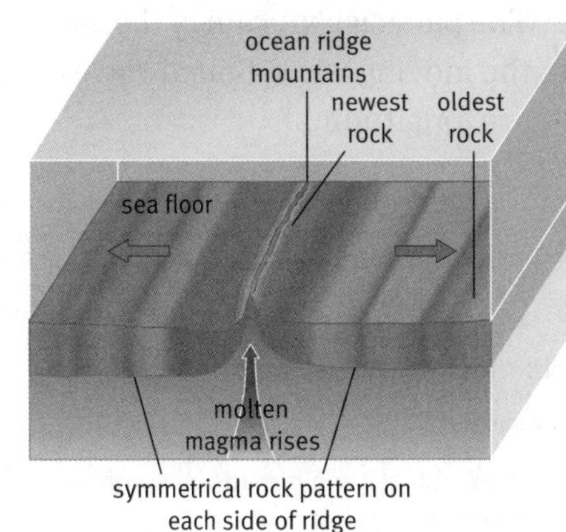

Seafloor spreading makes new rock.

P1.2.11–12 What is plate tectonics?

The outer layer of the Earth is made of about 12 huge pieces of rock, called **tectonic plates**. They move slowly all the time. Earthquakes, volcanoes, and mountain building usually happen where tectonic plates meet.

- In the Himalayas, plates move towards each other. They collide. The edges of the continents crumple together and pile up. This **builds mountains**.
- Most **volcanoes** are at plate boundaries where the crust is stretching or being compressed.
- In some places, tectonic plates slide past each other. Huge forces build up along fault lines. The forces become so big that the locked-together rocks break, and the plates move. The ground shakes. This is an **earthquake**.
- The movement of tectonic plates contributes to the rock cycle.

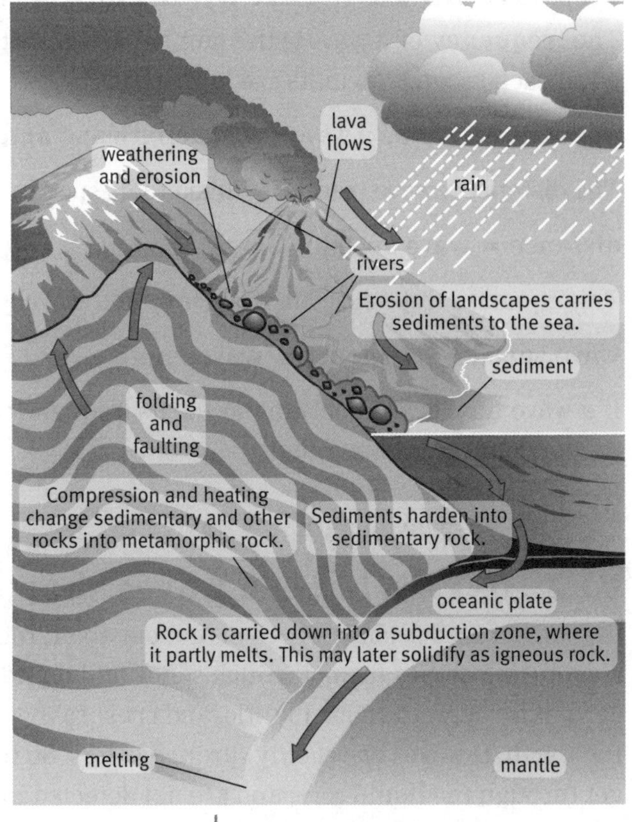

The movement of tectonic plates also plays a part in the rock cycle.

P1.2.13–14 What are seismic waves?

Earthquakes release energy. The energy spreads out as vibrations, or **seismic waves**. Seismometers detect waves on the Earth's surface. Two types of waves produced by earthquakes are:

- P-waves, which travel through solids and liquids
- S-waves, which travel through solids but not liquids.

P1.2.15, P1.2.19 What are waves?

A **wave** is a travelling vibration that transfers energy from place to place without transferring matter. There are two types of wave:

- **Longitudinal waves** travel as compressions. The particles vibrate in the same direction as the moving wave. Sound waves and P-waves are longitudinal.
- In a **transverse wave**, the particles vibrate at 90° to the direction of the wave's movement. Water waves and S-waves are transverse.

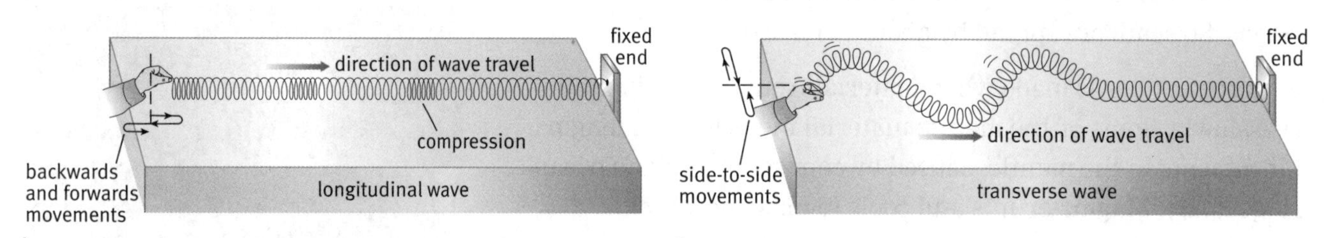

A longitudinal wave is made by compressing and releasing the spring.

A transverse wave is made by moving the spring from side to side.

P1.2.20–23 How can we measure waves?

The **frequency** of a wave is the number of waves that pass any point each second. Its units are hertz (Hz).

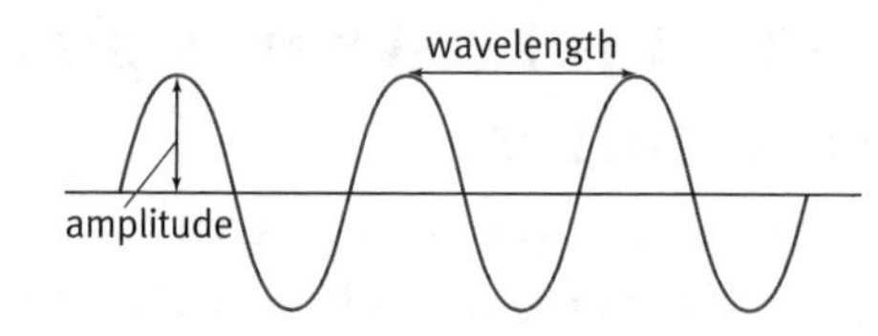

The diagram shows a wave's **wavelength** and **amplitude**.

The distance a wave travels is linked to its speed:

distance = wave speed × time

This equation links speed to frequency and wavelength:

wave speed = frequency × wavelength

If a wave travels at a constant speed, its frequency is inversely proportional to its wavelength:

$$\textbf{frequency} \propto \frac{\textbf{1}}{\textbf{wavelength}}$$

P1.2.16–17 What can waves show us?

Geologists used seismic waves to find out about the Earth's structure. P-waves travel through solids and liquids, so they can get through the core, mantle, and crust to the other side of the Earth. S-waves travel only through solids. So they cannot get through the liquid core, and are not detected in the S-wave **shadow zone**.

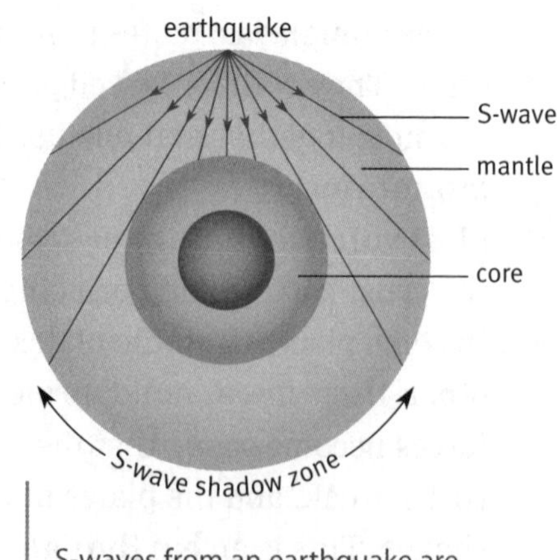

S-waves from an earthquake are blocked from reaching almost half of the Earth's surface.

Use extra paper to answer these questions if you need to.

1 Add labels to the diagram of the Earth to show the crust, mantle, and core.

2 Tick the boxes to show which objects below emit light.

- **a** the planet Jupiter ☐
- **b** the Sun ☐
- **c** the star Sirius ☐
- **d** the Moon ☐
- **e** comets ☐

3 Choose words from the box to fill in the gaps. The words may be used once, more than once, or not at all.

uncertainties	**assumptions**	
brightness	**parallax**	**pollution**

The distance to a star can be estimated from its relative __________ . It is difficult to make accurate observations of the night sky in cities because of light __________ . Even without this problem, there are __________ about the distances to stars. This is partly because scientists need to make __________ when interpreting observations.

4 Write definitions for the words below:

- **a** wave
- **b** frequency
- **c** wavelength
- **d** amplitude
- **e** light-year.

5 Calculate the distance travelled by each of the following waves:

- **a** An S-wave that travels at an average speed of 5 km/s for 410 seconds.
- **b** A P-wave that travels at an average speed of 10 km/s for 205 seconds.
- **c** A sound wave that travels at a speed of 340 m/s for 60 seconds.

6 The statements below describe how new rock is formed on the seafloor. Write the letters of the steps in the best order.

- **A** Some mantle material melts.
- **B** Magma erupts at the middle of the ridge.
- **C** Material from the solid mantle rises slowly.
- **D** This forms magma.
- **E** It cools to make new rock.

7 List three pieces of evidence Wegener used to support his theory of continental drift.

8 List three reasons to explain why Wegener's theory was not at first accepted by geologists.

9 Calculate the speed of each of the waves below.

- **a** a wave from an earthquake with a frequency of 0.5 Hz and a wavelength of 15 km
- **b** a wave from an earthquake with a frequency of 0.5 Hz and a wavelength of 18 km

10 The graph shows how the speeds of an S-wave and a P-wave change at they travel through the Earth.

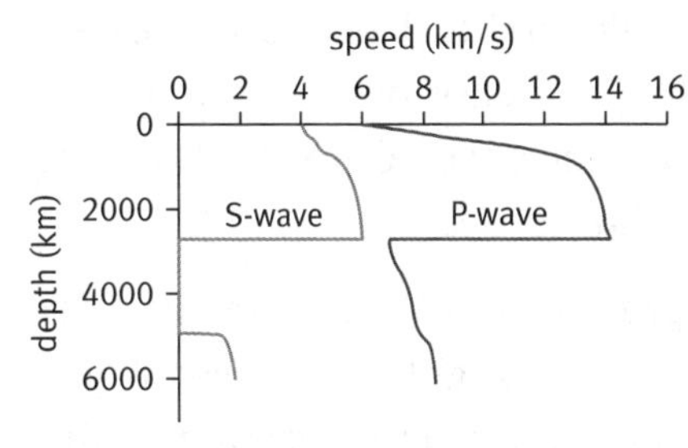

- **a** Use the graph to estimate the depth at which the Earth's solid mantle meets the liquid outer core.
- **b** Describe the evidence on the graph that supports the theory that Earth has a liquid outer core.

H **11** Describe evidence for the explanations below:

- **a** Distant galaxies are moving away from us.
- **b** Space itself is expanding.
- **c** The Earth's magnetic field can change direction.

12 Annotate the diagram below to show how the movement of tectonic plates contributes to the rock cycle.

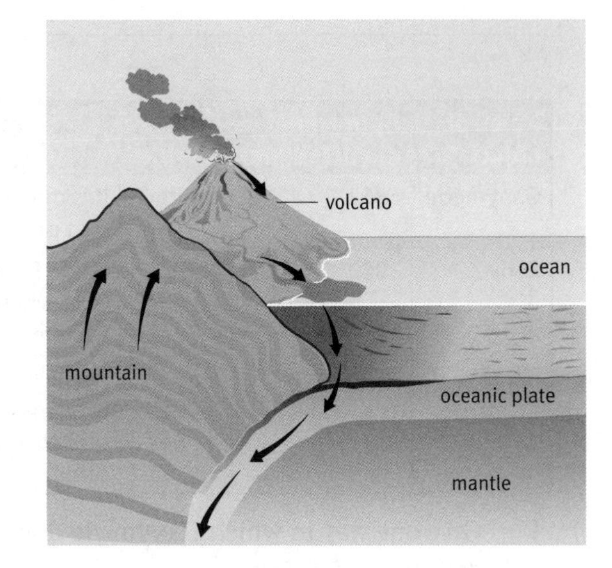

P1

Who first saw Ganymede?
Many astronomers accept that Galileo was the first person to see Jupiter's moon Ganymede. Galileo used a telescope to make his observations in 1610.

However, there is evidence that a Chinese astronomer, Gan Dej, first saw Ganymede about 2000 years earlier. Gan Dej reported a 'small reddish star' next to Jupiter. This could have been the moon we now call Ganymede. However, Gan Dej did not have a telescope. The moons of Jupiter are not bright enough for their colours to be seen without a telescope.

1 Read the article.

a Read the following statements.

Put a tick (√) in the box next to each of the three correct statements.

Galileo saw Ganymede through a telescope. ☐
Ganymede is a star. ☐
Gan Dej saw an object close to Jupiter without a telescope. ☐
Ganymede is a moon. ☐
Gan Dej definitely saw Ganymede. ☐
Galileo was definitely the first person to see Ganymede. ☐ [3]

b Use the information below, and the data in the table, to evaluate whether or not it might have been possible for Gan Dej to have seen Ganymede 2000 years before Galileo, without a telescope.

The *apparent magnitude* of an object in Space is a measure of its brightness as seen from Earth. The table gives some values for apparent magnitude. The smaller the apparent magnitude, the brighter the object.

Apparent magnitude	Object
between 3 and 4	Faintest object in Space that can be seen from a modern city without a telescope .
4.4	Ganymede
6.5	Faintest object in Space that can be seen in a very dark sky without a telescope

____

____ [2]

c

Name of object	Approximate diameter (km)	Object it orbits	Other information
Ganymede	5362	Jupiter	It is one of many objects that orbits Jupiter.
Xena	3000	Sun	Made from gases and dust when the Solar System began.
Pluto	2400	Sun	Its mass is less than the total mass of the other objects that cross its orbit.

In August 2006, the International Astronomical Union defined a planet as a body that orbits a star, and has a spherical shape. The mass of a planet must be much greater than the total mass of the other objects that cross its orbit.

i Give one way in which Ganymede **does not** fit the definition of a planet.

____ [1]

ii Give one way in which Pluto **does not** fit the definition of a planet.

____________________________________ [1]

iii Give one way in which Xena **does** fit the definition of a planet.

____________________________________ [1]

Total [8]

2 a On the diagram of the Earth, label the **crust, mantle,** and **core.** [2]

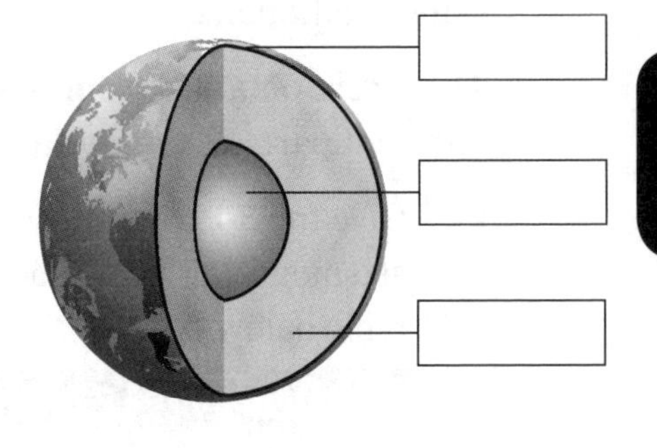

b Draw straight lines to match each **explanation** with **evidence** that supports it.

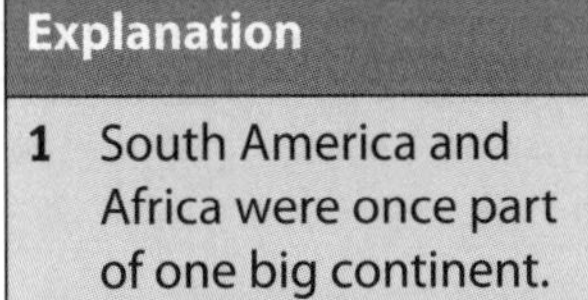

Explanation
1 South America and Africa were once part of one big continent.
2 Mountains are being formed all the time.
3 The Earth is older than 4000 million years.
4 Asteroids have collided with Earth.

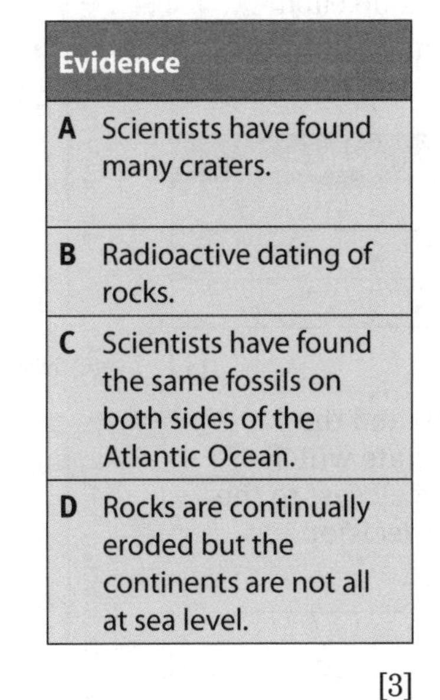

Evidence
A Scientists have found many craters.
B Radioactive dating of rocks.
C Scientists have found the same fossils on both sides of the Atlantic Ocean.
D Rocks are continually eroded but the continents are not all at sea level.

[3]

Total [5]

3 A seismometer detects seismic waves from earthquakes. It records data on seismographs like the one below.

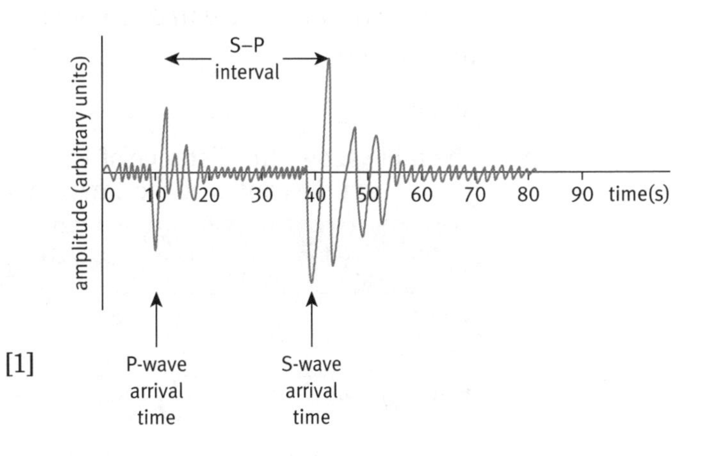

a Use the graph on the right to decide which travel faster: S-waves or P-waves. Explain your decision.

____

____________________________________ [1]

b

> On a seismograph, scientists use the S-P interval to calculate the distance from the centre of the earthquake to the seismometer. The S-P interval is marked on the seismograph on the previous page. The shorter the S-P interval, the closer the seismometer was to the earthquake.

The seismographs on the right are from the same earthquake.

They were obtained from seismometers in different parts of the world.

i Give the S-P interval shown on each seismograph. Write your answers in the table below.

seismograph	S-P time interval (s)
A	
B	
C	

[1]

ii Use your answer to part (i) and the information in the box to state which seismograph was recorded closest to the earthquake. Explain your decision.

____

______________________________ [1]

c i Assume the S waves from the earthquake travelled at a speed of 5 km/s. Calculate the distance they travelled in 60 seconds. Show your working.

Answer = ________ km [2]

ii The table shows the distance of three towns from the centre of the earthquake.

Town	distance from centre of earthquake (km)
Owlton	250
Foxford	670
Badgerbridge	290

Which town or towns would S-waves from the earthquake have reached in 60 seconds or less?

______________________________ [1]

Total [6]

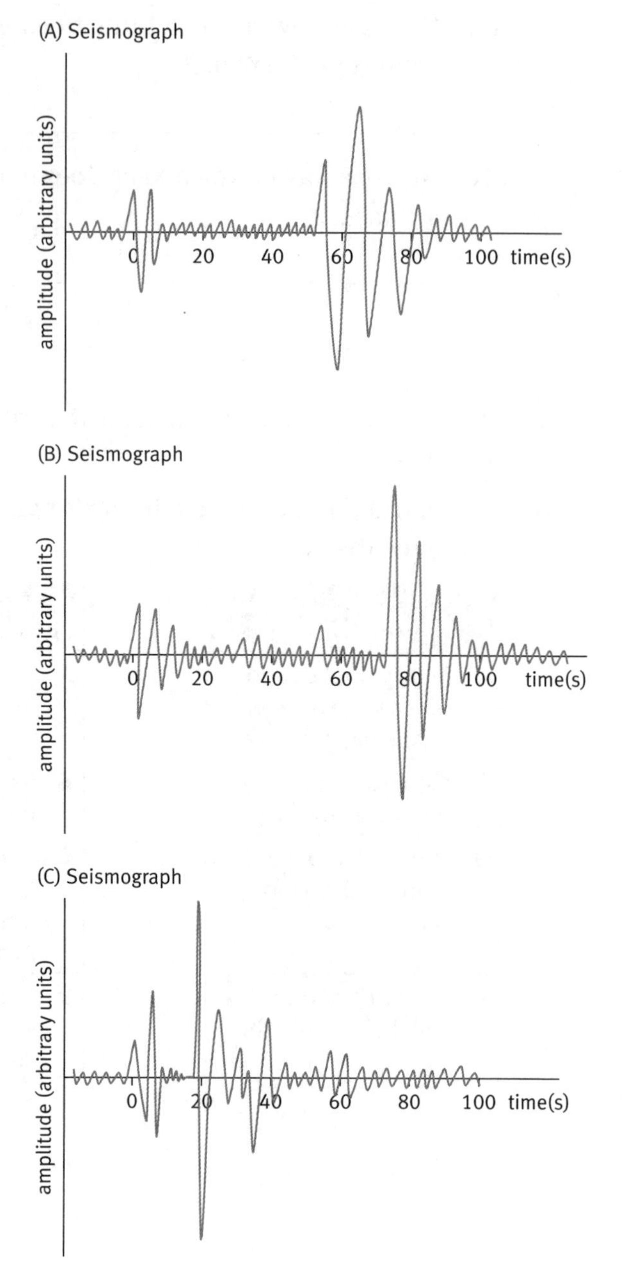

Exam tip

Make sure you know the differences between S-waves and P-waves.

P2 Radiation and life

1 Use the words in the box to finish labelling the diagram.

detector	**energy of each photon**	**intensity**	**absorbs**
number of photons	**transmits**	**reflects**	**source**

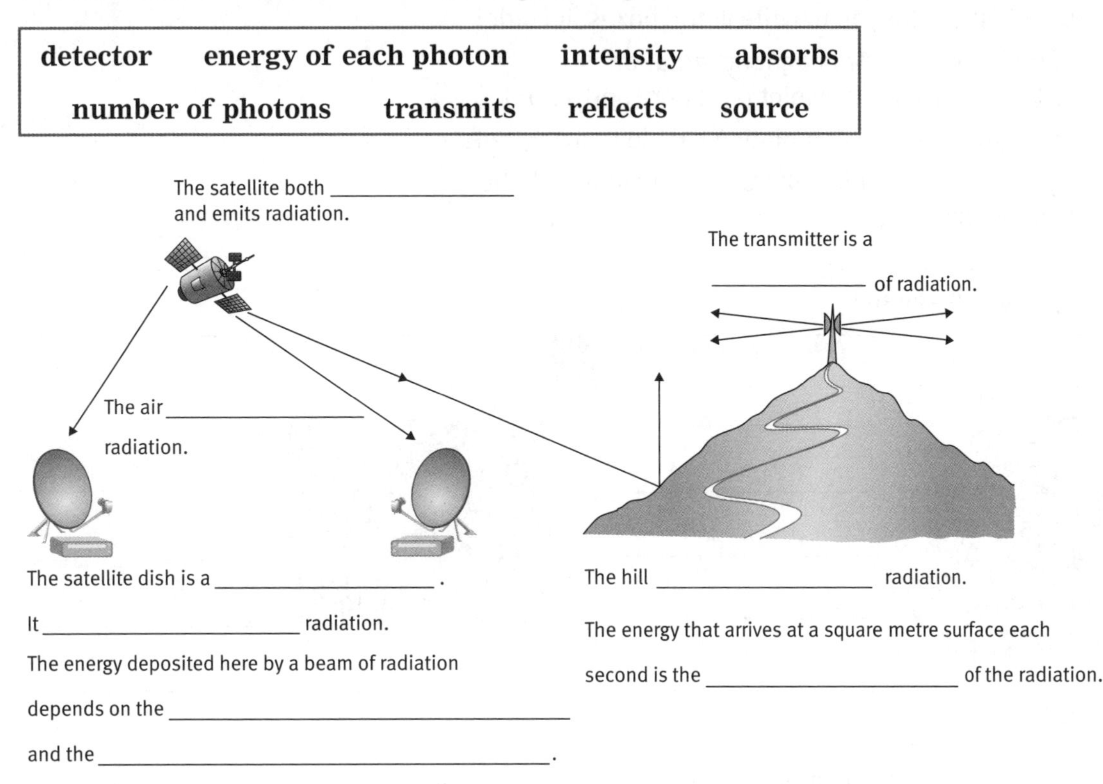

The satellite dish is a ______________ .

It ______________ radiation.

The energy deposited here by a beam of radiation depends on the ______________

and the ______________ .

The hill ______________ radiation.

The energy that arrives at a square metre surface each second is the ______________ of the radiation.

2 Solve the clues to fill in the grid.

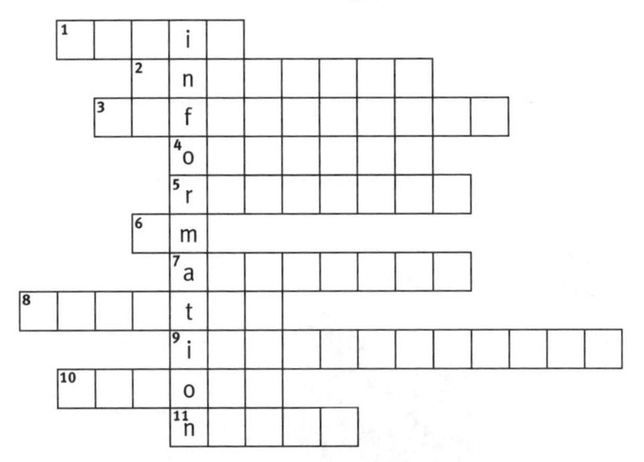

1 Signals are carried through space and the Earth's atmosphere by microwaves and ______ waves.

2 Signals are carried through optical fibres by light waves and ______ rays.

3 The more ______ stored, the better the quality of an image or sound.

4 Light and infrared carry information along ______ fibres.

5 A radio ______ decodes a radio wave's pattern of variation to reproduce the original sound.

6 The signal carried by FM and ______ radio waves varies in exactly the same way as the information from the original sound wave.

7 The signal carried by FM waves is called an ______ signal.

8 Sound waves can be converted into a ______ code made of two signals.

9 Analogue and digital signals pick up random unwanted signals as they travel. This is called ______, or noise.

10 Digital radio receivers pick up pulses and ______ them to make a copy of the original sound wave.

11 Digital radio receivers clean up signals to remove ______.

3 Add captions to each picture. Use a separate sheet of paper if there is not enough space for you to write in the boxes. Include:

- the name of the type of electromagnetic radiation represented (radio waves, ultraviolet radiation, and so on)
- the damage (if any) this type of radiation can do to living cells
- what Alex can do to protect himself from this type of radiation (if he needs to do anything).

Alex's holiday: a day in the life
1
2
3
Is that the dentist? My filling's just fallen out.
4
5
6
7

P2.1.1 What is electromagnetic radiation?

Electromagnetic radiation travels as waves. It carries energy.

- A **source** gives out radiation.
- The radiation spreads out from its source'
- The radiation may be **reflected**, **transmitted**, or **absorbed**.
- Some of the radiation may be absorbed by a **detector**.

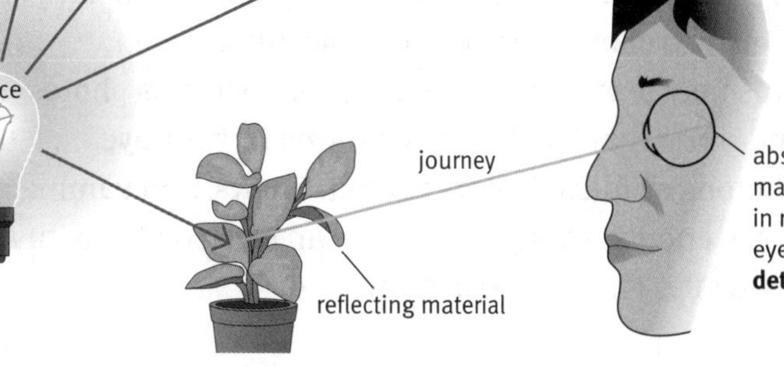

The radiation is transmitted through the air, reflected (and absorbed) by the plant, and absorbed by the eye.

P2.1.3, P2.1.7–9 How much energy does radiation transfer?

Electromagnetic radiation transfers energy in packets, or **photons**. The amount of energy absorbed by a detector depends on:

- the number of photons
- the amount of energy each photon carries.

The amount of energy that arrives at a square metre of a surface each second is the **intensity** of the radiation.

As it gets further from its source, the intensity of a beam of electromagnetic radiation decreases.

H This is because, as it spreads out, the radiation reaches bigger and bigger surface areas. Also, some of the radiation is absorbed by the medium it is travelling through.

P2.1.2, P2.1.4–6 What is the electromagnetic spectrum?

The electromagnetic spectrum includes these radiations:

radio waves — microwaves — infrared — visible light — ultraviolet — X-rays — gamma rays →

increasing frequency

Exam tip

Learn the order of the radiations in the diagram – you might be asked about it in an exam.

All types of electromagnetic radiation travel at 300 000 km/s through a vacuum. The photons of high frequency waves carry more energy than the photons of low frequency waves.

P2.1.10–12 What happens when materials absorb radiation?

When materials absorb electromagnetic radiation, they gain energy.

- Radio waves and microwaves cause a varying electric current in a metal wire, such as a radio or phone aerial.

P2

- Microwaves and infrared waves heat materials.
- Ultraviolet radiation, X-rays, and gamma rays have enough energy to change atoms and molecules.
  This may lead to chemical reactions, such as photosynthesis and those that happen in the retina of your eye.
- Photons of ultraviolet radiation, X-rays, and gamma rays can carry enough energy to remove an electron from an atom or molecule. This is **ionisation**.

P2.2.1–4 Do microwaves harm humans?

Microwaves, light, and infrared radiation can heat materials by making their particles vibrate faster. This **heating effect** depends on the radiation's intensity and the length of time it is absorbed.

The heating effect can damage cells. Some people think that low intensity microwave radiation from mobile phone handsets is a risk to health. Others say the evidence does not support this claim.

Some microwaves are strongly absorbed by water. They are used for cooking water-containing foods in microwave ovens. Microwave ovens have metal cases and door screens that reflect or absorb microwaves to stop them leaving the oven.

P2.2.6–7 What damage do gamma rays do?

Radioactive materials emit gamma radiation all the time. This ionising radiation damages the DNA of living cells. The damage may lead to cancer or cell death.

P2.2.8–10 What protects us from UV?

The Sun emits ultraviolet radiation (UV). This can cause skin cancer.

A layer of gases in the atmosphere – the **ozone layer** – protects living organisms by absorbing most of the UV radiation from the Sun. People use sun-screens and clothes to absorb harmful UV radiation that gets through the ozone layer.

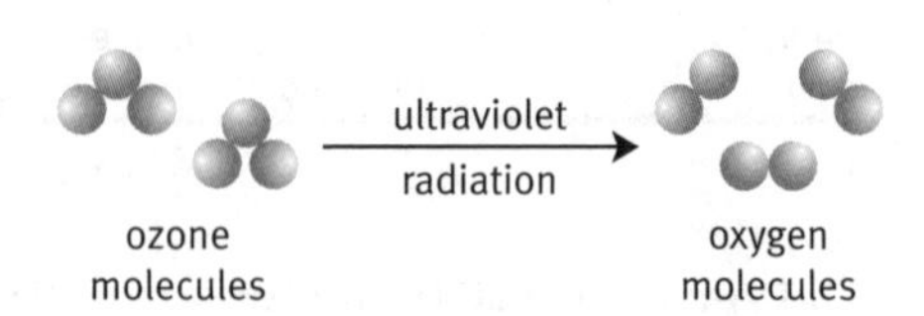

When ozone absorbs UV radiation, its molecules break down.

P2.2.11 How are X-rays useful?

X-rays are absorbed by dense materials, but pass through less dense ones. This means they are used to produce:

- shadow pictures of bones in our bodies
- shadow pictures of objects in luggage at airports.

People who work with X-rays are protected from radiation by dense materials such as lead and concrete.

P2.3.1–4 How does electromagnetic radiation make life possible?

All objects emit electromagnetic radiation. The **principal frequency** of the radiation emitted by an object is the frequency that is emitted with the highest intensity. The higher the temperature of an object, the higher its principal frequency.

P2.3.5–11 What is the greenhouse effect?

Greenhouse gases keep the Earth warmer than it would otherwise be. There are three main greenhouse gases in the atmosphere:

- carbon dioxide (the most important greenhouse gas, present in small amounts compared to other atmospheric gases)
- methane (trace amounts)
- water vapour.

The concentration of carbon dioxide in the atmosphere hardly changed for thousands of years. The carbon dioxide removed by photosynthesis was balanced by that returned by respiration. There is a diagram of the carbon cycle on page 74.

Since 1800 the concentration of carbon dioxide increased, mainly because humans:

- burn increasing amounts of fossil fuels
- cut down and burn forests to clear land.

H Computer climate models provide evidence that human activities are causing global warming.

Global warming may lead to:

- climate change, meaning that some food crops will no longer grow in some places
- ice melting and seawater expanding as it warms up, causing rising sea levels and flooding of low-lying land
- more extreme weather conditions

  H because higher temperatures cause more convection in the atmosphere, and more evaporation of water from oceans and the land.

P2.4.1–2 How does radiation carry information?

Electromagnetic waves travel from a source to a detector. We can use them to transmit information.

- Microwaves, and some radio waves, are not strongly absorbed by the atmosphere. They carry information for radio and TV programmes through the air.
- Visible light and infrared radiation are not absorbed by glass. They carry information along optical fibres for cable TV and high-speed Internet connections.

We send information from place to place using a **carrier wave**. The carrier wave is the radio wave, microwave, visible light, or infrared radiation. Adding information to the carrier wave creates a signal.

P2.4.3–7 Analogue or digital?

An **analogue signal** changes all the time. For example FM and AM radio waves vary in the same way as the information from the original sound wave.

A **digital signal** can take just two values. Its code is made up of two symbols, 0 and 1. The coded information is carried by switching the carrier wave on and off. This makes short bursts of waves, or **pulses**:

- 0 = no pulse
- 1 = pulse.

When a radio, mobile phone, or computer receives the waves, a processor in the device decodes the pulses. This converts the digital signal back to the original analogue signal.

The advantages of digital signals include:

- the information can be stored and processed by microprocessors in computers and phones
- digital information can be stored in small memories.

As analogue and digital signals travel, they pick up unwanted electrical signals. This is **noise**, or **interference**. It is easier to remove noise from a digital signal than an analogue one.

H This is because, for digital signals, 0 and 1 can still be recognised if noise has been picked up. The signal can be 'cleaned up' by removing the noise. Noise cannot be removed from analogue signals.

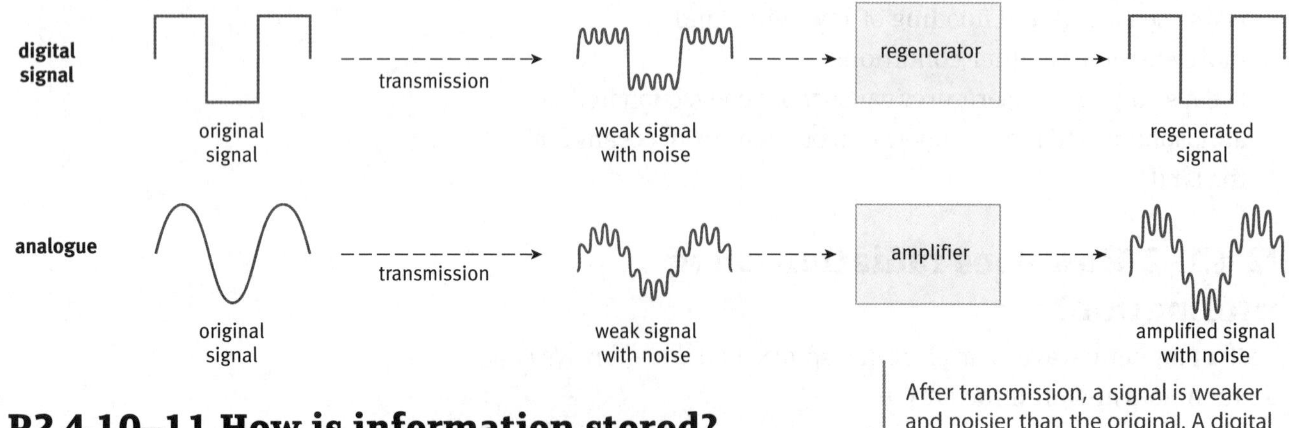

After transmission, a signal is weaker and noisier than the original. A digital signal can be 'cleaned up' by a regenerator.

P2.4.10–11 How is information stored?

The amount of information needed to store an image or sound is measured in bytes (B). The more information stored, the better the quality of the image or sound.

Use extra paper to answer these questions if you need to.

1 Write each type of electromagnetic radiation below in the correct column of a copy of the table.

a high energy ultraviolet
b light
c X-rays
d infrared
e gamma rays
f microwaves

Ionising radiations	Radiations that cause a heating effect only

2 Write the letters of the radiations below in order of increasing frequency.

A X-rays
B infrared
C visible
D microwaves
E gamma rays
F ultraviolet
G radio waves

3 Write the letter C next to the statements that are true for carbon dioxide. Write the letter O next to the statements that are true for ozone. Write B next to statements that are true for both.

a This gas is added to the atmosphere by respiration.
b This gas absorbs ultraviolet radiation.
c This gas is present in the Earth's atmosphere.
d This gas helps to prevent people getting skin cancer.
e This gas is removed from the atmosphere by photosynthesis.
f The amount of this gas in the atmosphere is increasing.

4 Highlight the statements below that are **true**. Write a correct version of the false statement.

a The higher the frequency of an electromagnetic radiation, the less energy is transferred by each photon.
b Ionising radiation removes electrons from atoms or molecules.
c Metals reflect microwaves.

5 Use the words in the box to fill in the gaps. Each word can be used once, more than once, or not at all.

radio **light** **absorbed** **reflected** **TV**
microwaves **infrared** **transmitted**

Radio waves and _______ carry information for _______ and _______ because they are not strongly _______ by the atmosphere. _______ and _______ carry information along optical fibres because the radiation is not _______ much by glass.

6 Which picture (X, Y, or Z) has the best quality image? Explain how you know. All three pictures are printed the same size.

Picture	Amount of information
X	1 Mb
Y	100 kB
Z	10 kB

7 Draw lines to match each wave to its description.

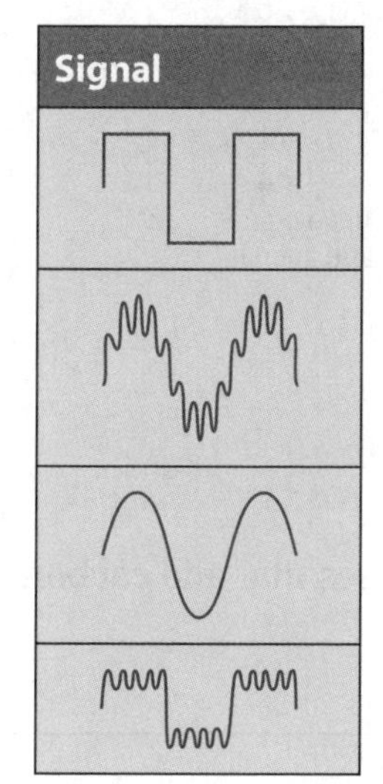

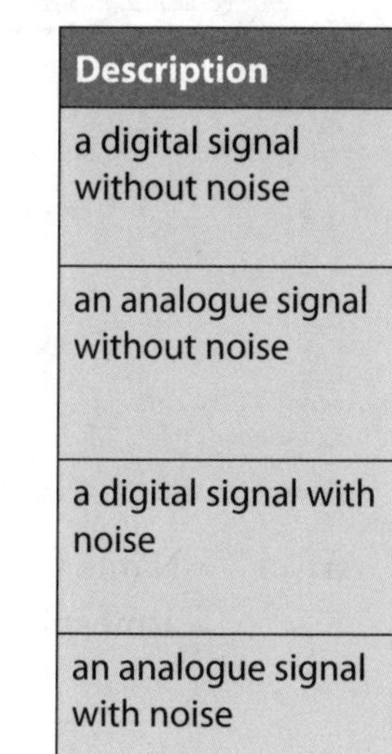

Description
a digital signal without noise
an analogue signal without noise
a digital signal with noise
an analogue signal with noise

8 This question is about the greenhouse effect.

a Name two natural processes that add carbon dioxide to the atmosphere.
b Name one natural process that removes carbon dioxide from the atmosphere.
c Explain why the amount of carbon dioxide in the atmosphere has been increasing over the past 200 years.
d Name two greenhouse gases, other than carbon dioxide.
e List three possible effects of global warming.

9 Name substances that absorb each of the following types of radiation:

a microwaves
b ultraviolet radiation
c X-rays.

H **10** Write a definition for the **intensity** of a beam of electromagnetic radiation.

11 Explain why the intensity of a beam of electromagnetic radiation decreases with distance from the source.

12 Write the formula of an ozone molecule and an oxygen molecule. Describe what happens when an ozone molecule absorbs ultraviolet radiation.

13 Explain what the term **principal frequency** means. Which has a higher principal frequency – the Earth or the Sun?

14 Explain why global warming could result in more extreme weather events.

1 The diagram shows part of the carbon cycle.

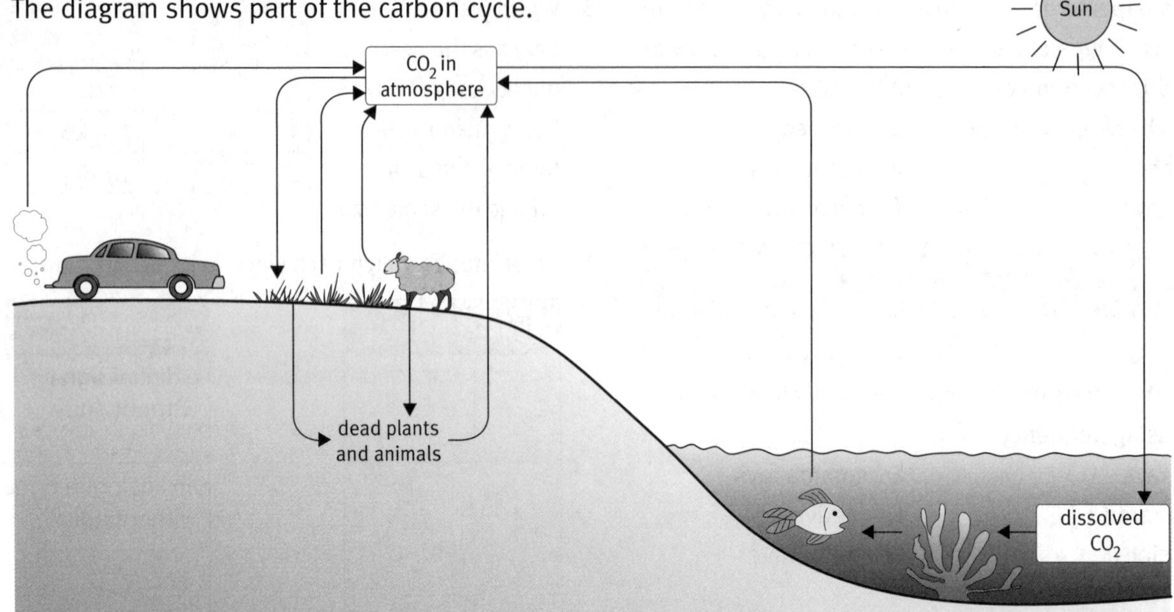

a **i** Name two processes that add carbon dioxide to the atmosphere.

______________________________________ [2]

ii Name two processes that remove carbon dioxide from the atmosphere.

______________________________________ [2]

b Increasing amounts of carbon dioxide in the atmosphere cause global warming.

i Give two problems caused by global warming.

____

______________________________________ [2]

H **ii** Name two greenhouse gases other than carbon dioxide.

______________________________________ [2]

Total [8]

2

Scientists have invented a scanner to find out if premature babies are at risk of brain damage.

The scanner sends beams of light into the brain. It uses light of two wavelengths: 780 nm and 815 nm.

Some of the light passes through brain tissue. Some of the light is absorbed by water in the brain. Most of the light is scattered in all directions.

Detectors in the scanner measure the intensity of the light that comes out of the brain. If the intensity is less than expected, the baby's brain might be bleeding.

The scanner builds up a 3-dimensional image of the brain.

Doctors can use this to find out where the bleeding is.

a What scientific word means that light **passes through** brain tissue?

Draw a (ring) around the correct answer.

transmitted **reflected** **absorbed** [1]

b Use the information in the box to decide whether blood transmits, reflects, or absorbs the light that the scanner emits.

Draw a (ring) around the correct answer.

transmits **reflects** **absorbs** [1]

Give a reason for your decision.

____

____

____

______________________________________ [2]

H **c** The diagram shows some of the detectors around a baby's head.

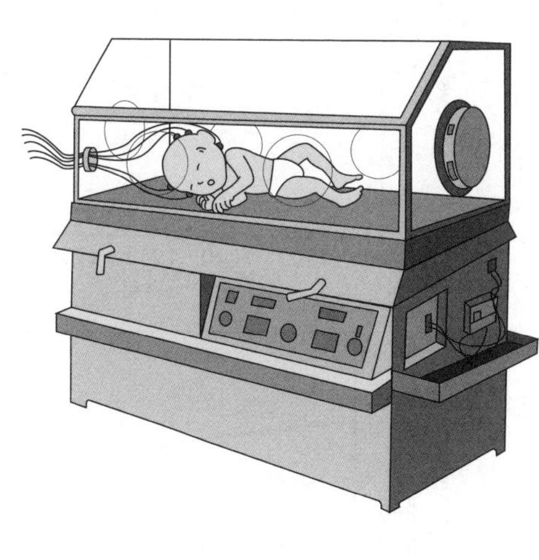

Each detector records a different reading for light intensity.
Why are the readings different?
Tick the **best** answer.

The amount of energy carried by a photon does not change. ☐

In one second, a different number of photons arrives at each detector. ☐

The amount of energy carried by a photon changes each second. ☐

Each photon carries the same amount of energy. ☐ [1]

d i Give one reason why it would not be sensible for the scanner to send X-rays into the brain.

______________________________________ [1]

ii Give one reason why it would not be sensible for the scanner to send microwaves into the brain.

______________________________________ [1]

Total [7]

3 Methane and carbon dioxide are greenhouse gases.

The quality of written communication will be assessed in your answer to this question.

Write your answer on separate paper or in your exercise book. Use evidence from the five graphs below to identify and evaluate evidence for global warming, and its causes and effects.

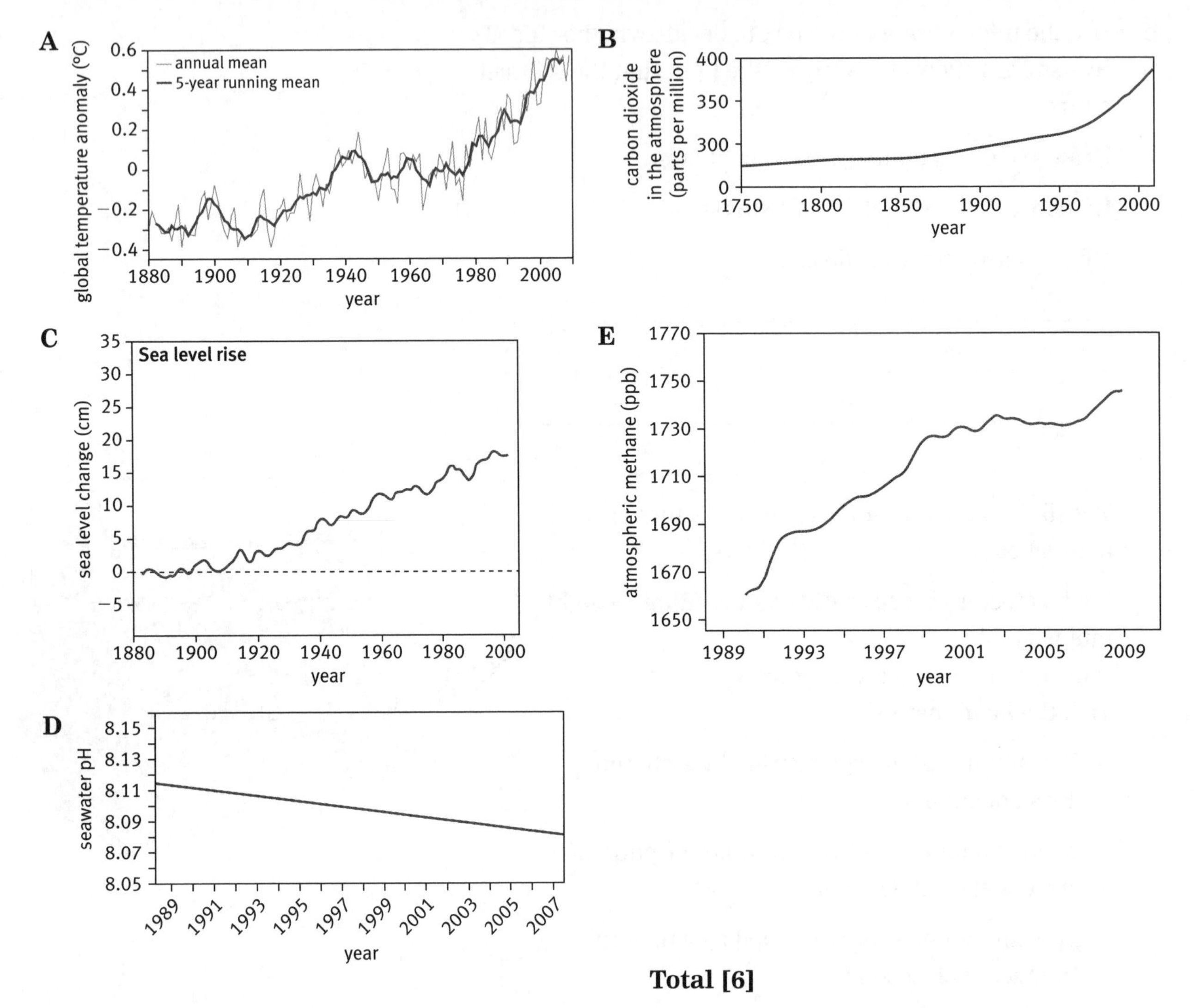

Total [6]

4 Beams of electromagnetic radiation from the phone mast deliver photons ('packets') of energy to Mike's and Helen's mobile phones.

The amount of energy delivered by each photon is the same. Explain why the amount of energy that arrives at Helen's mobile phone is less than the amount of energy that arrives at Mike's mobile phone.

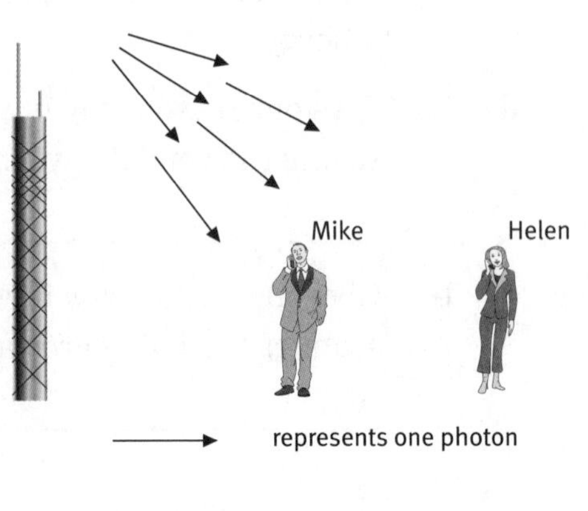

______________________________ **[1]**

Total [1]

P3 Sustainable energy workout

1 Write each letter in an appropriate box to summarise some different ways of generating electricity.

A heat up water to make steam

B wave movement

C tidal movement

D generator – a big coil of wire turns in a magnetic field

E solar voltaic cells

F releases carbon dioxide gas

G wind movement

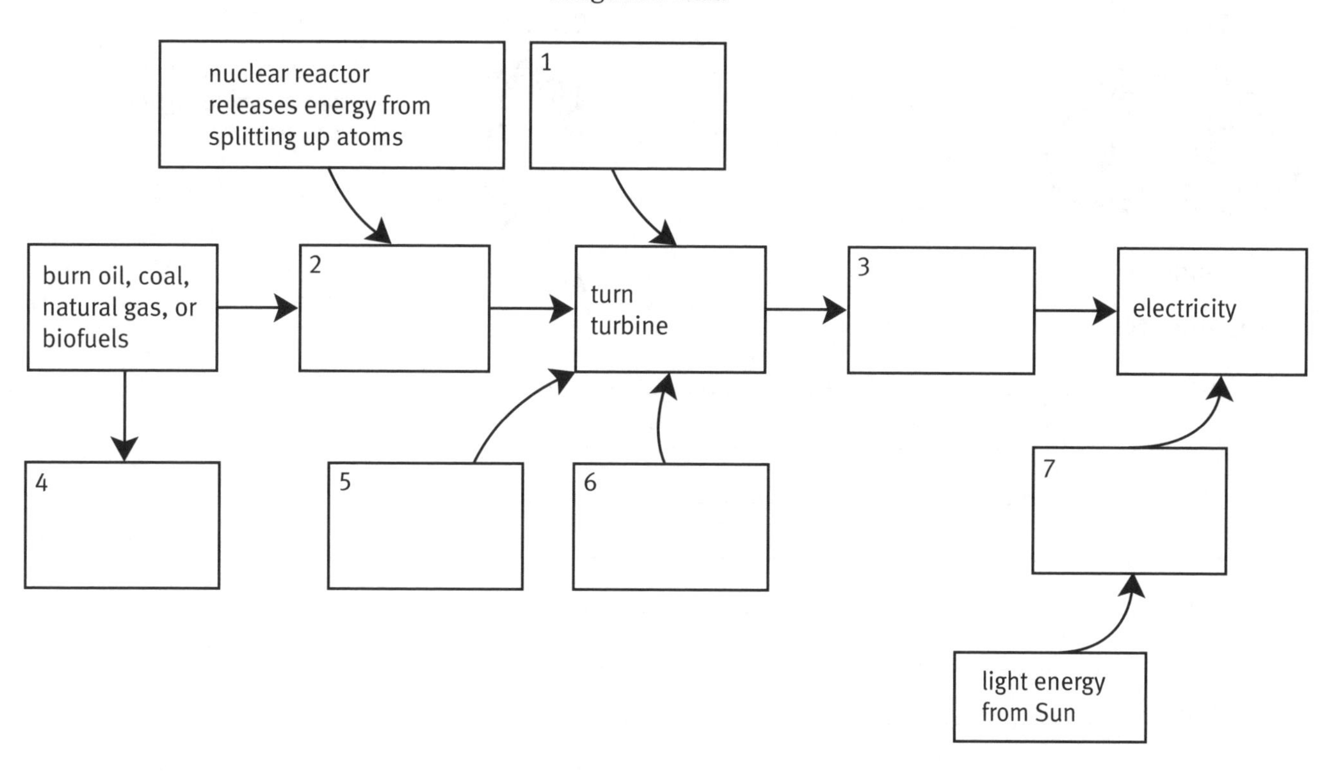

2 The data in the table are about electricity generated from different energy sources.

Use the data to write one argument that supports each person's opinion.

	Nuclear	Wind	Coal
Approximate efficiency	35%	59% of wind's energy can be extracted by blades	35%
Environmental impact	produce radioactive waste	some people think they look unattractive and that they are too noisy	contribute to acid rain
Cost per unit of electricity (pence)	3.0–4.0	3.0–4.0	3.0–3.5
Tonnes of carbon dioxide made for one terajoule of electricity	30	10	260

Opinions:

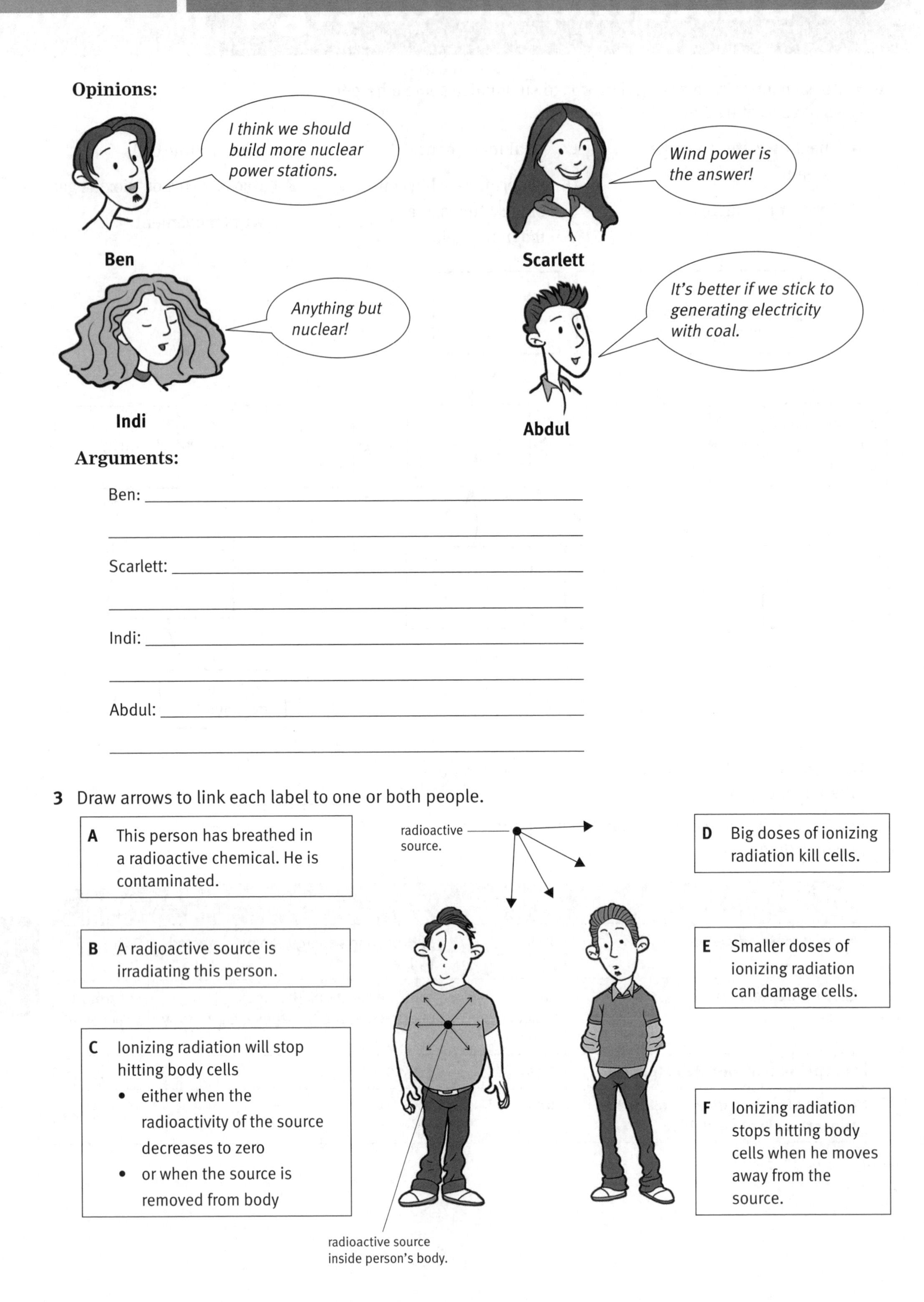

I think we should build more nuclear power stations.
Ben
Wind power is the answer!
Scarlett
Anything but nuclear!
Indi
It's better if we stick to generating electricity with coal.
Abdul
radioactive source.
radioactive source inside person's body.

Arguments:

Ben: ______________________________

____

Scarlett: ______________________________

____

Indi: ______________________________

____

Abdul: ______________________________

____

3 Draw arrows to link each label to one or both people.

A This person has breathed in a radioactive chemical. He is contaminated.

B A radioactive source is irradiating this person.

C Ionizing radiation will stop hitting body cells
- either when the radioactivity of the source decreases to zero
- or when the source is removed from body

D Big doses of ionizing radiation kill cells.

E Smaller doses of ionizing radiation can damage cells.

F Ionizing radiation stops hitting body cells when he moves away from the source.

P3.1.1–3 What energy sources do we use?

The demand for energy is increasing because:

- the world's population is increasing
- people travel further and have more possessions.

The increasing energy demand causes concerns about:

- the availability of energy sources
- the environmental impacts of using energy sources.

Primary energy sources exist naturally. They include fossil fuels (coal, oil, and gas), nuclear fuels, biofuels, wind, waves, and radiation from the Sun.

Electricity is a **secondary energy source**. It is generated from a primary source.

P3.1.5–8 How do we calculate electricity use?

When an electric current passes through a device, energy is transferred from the power supply to the device and the environment.

The power of a device or power station is the amount of energy it transfers each second,

H or the rate at which it transfers energy. Different devices have different power ratings.

You can use this equation to calculate the power of a device:

power	=	**voltage**	×	**current**
(watts, W)		(volts, V)		(amps, A)

Use this equation to calculate the energy transferred by a device:

energy transferred	=	**power**	×	**time**
(joules, J)		(watts, W)		(seconds, s)
(kilowatt-hours, kWh)		(kilowatts, kW)		(hours, h)

A joule is a tiny amount of energy, so home electricity meters measure energy transfer in **kilowatt-hours**. One kilowatt-hour is the energy transferred by a 1 kW appliance in 1 hour.

3 600 000 J = 1 kWh = 1 **unit**

You can calculate the cost of energy supplied by electricity like this:

cost = power × time × cost per kilowatt-hour

Worked example

Jason spends half an hour ironing shirts with a 3 kW iron. One unit of electricity costs 10p.

cost = 3 kW × 0.5 h × 10p/kWh = 15p

P3.1.11,14, P3.3.1 Can we cut energy use?

Individuals use energy for heating, transport, and using electrical devices. Energy is used to produce food, drink, and the other things we buy.

Public services and companies use energy for activities like building houses, running supermarkets, and powering computer servers. Individuals and workplaces can cut energy use by:

- switching off electrical devices
- using devices that are energy efficient.

Nationally, we could cut energy use by improving public transport, or by generating electricity more efficiently.

P3.1.12–13 What is efficiency?

For an electrical device or a power station:

$$\textbf{efficiency} = \frac{\textbf{energy usefully transferred}}{\textbf{total energy supplied}} \times 100\%$$

Worked example

A 150 W flat screen TV transfers 50 J of energy as light and sound each second.

In one second, the total energy supplied is 150 J.

So the efficiency of the TV is $\frac{50}{150} \times 100\% = 33\%$

The Sankey diagram shows how energy is transferred by the TV. The total width of arrows on a Sankey diagram stays the same, because energy is **conserved** – it cannot be created or destroyed.

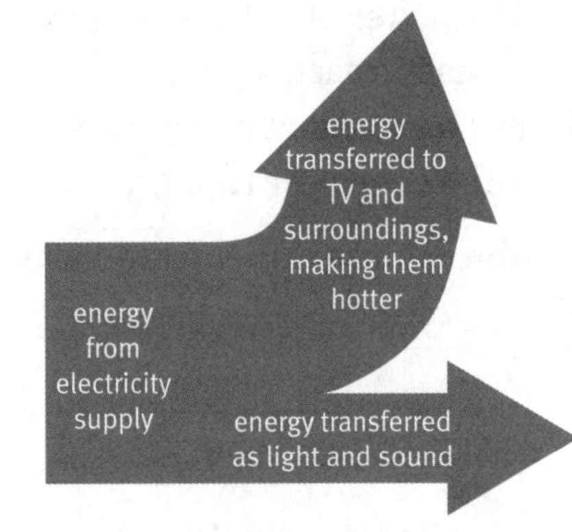

P3.2.1–6 How is electricity generated?

Electricity is convenient. It can be used in many ways, and transmitted long distances.

Generators make electricity by **electromagnetic induction**. In a generator, a magnet spins near a coil of wire. This induces a voltage across the ends of the coil. If the coil is part of a circuit, the induced voltage makes a current flow.

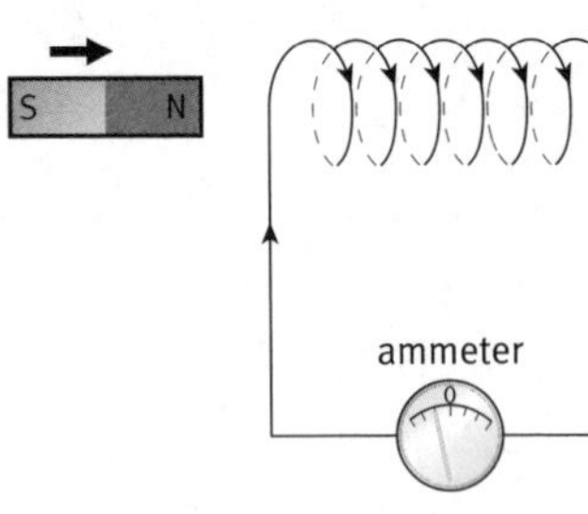

While the bar magnet is moving into the coil, there is a small reading on the sensitive ammeter.

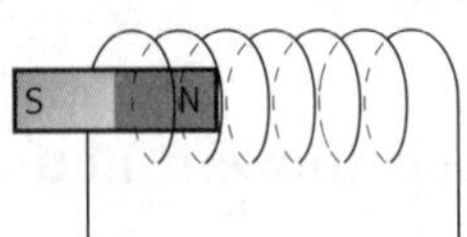
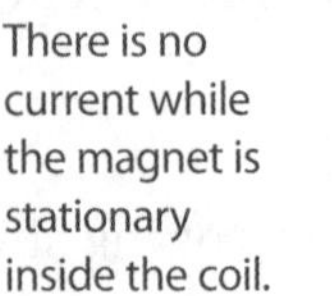

There is no current while the magnet is stationary inside the coil.

While the magnet is being removed from the coil, there is again a small current, but now in the opposite direction.

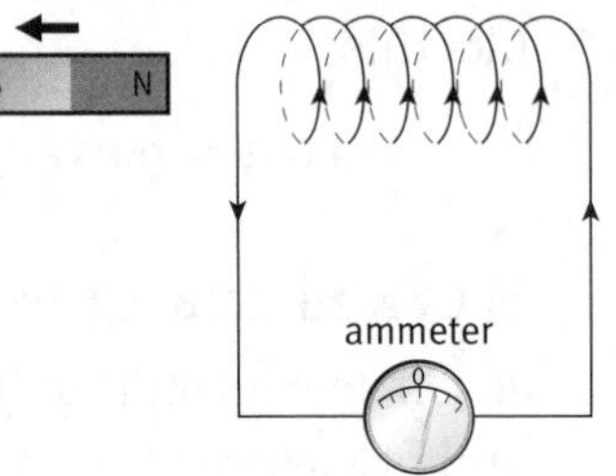

In **thermal power stations**, steam keeps the coil spinning. The steam comes from heating water. Primary sources (fossil fuels, nuclear power, or biofuels) supply the heat.

The more primary fuel supplied each second, the greater the current produced.

The diagram shows how thermal power stations generate electricity.

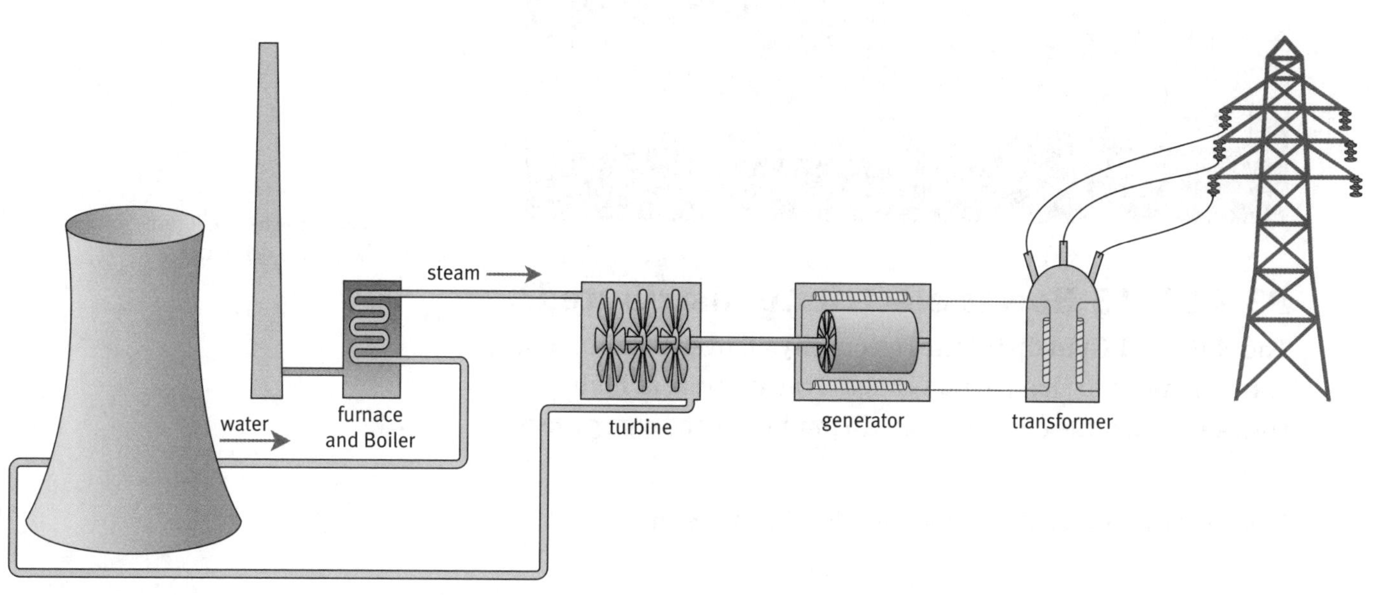

Electrical generation is never 100% efficient. The Sankey diagrams show the efficiency of coal-fired and gas-fired power stations.

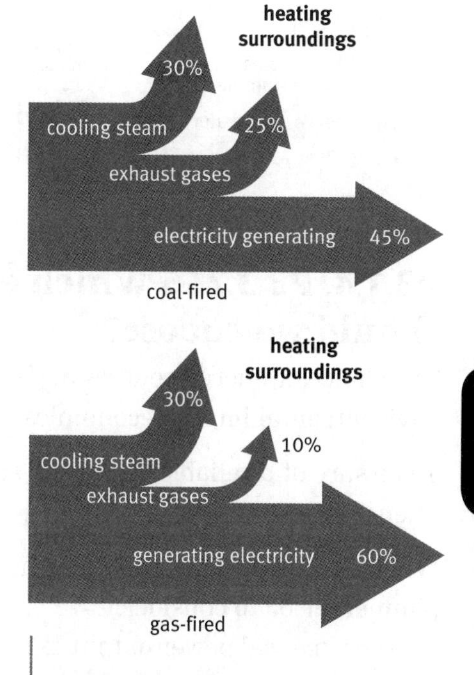

Sankey diagrams show what happens to all the energy. Less energy is wasted in a gas-fired power station.

P3.2.7–10 What happens in nuclear power stations?

In nuclear reactors, uranium atoms split up. This releases heat. The heat boils water to make steam. The steam turns a turbine connected to a generator, as in a thermal power station.

Nuclear fuels are **radioactive**. They make **radioactive waste**. Radioactive waste emits **ionising radiation**, which damages living cells. This damage can lead to cancer or cell death.

If you are exposed to ionising radiation, you have been **irradiated**. Radioactive **contamination** happens when radioactive material lands on or in a person or object. The person or object becomes radioactive, which may lead to long-term irradiation.

P3.2.11 How do renewable sources generate electricity?

Many renewable energy sources turn turbines directly.

- Moving air turns **wind** turbines.
- **Wave** movements turn wave turbines.
- Falling **water** turns turbines in **hydroelectric** power stations.

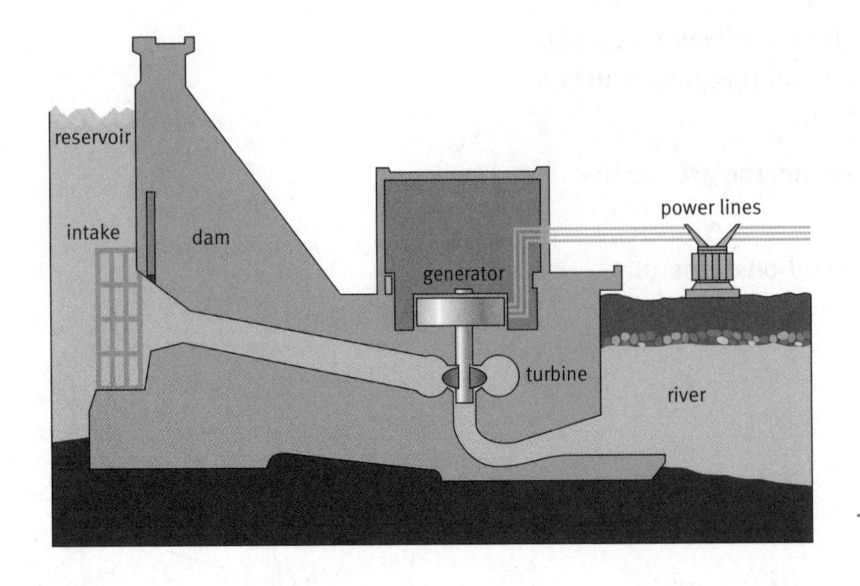

Water from the reservoir turns turbines, which turn the generator.

P3.2.12–13 How is electricity distributed?

The **National Grid** distributes electricity all over Britain. It uses **transformers** to change the voltage. Transmitting at high voltage (so that the current is small) minimises the energy lost as heat.

Electricity is supplied to homes at 230 volts in the UK.

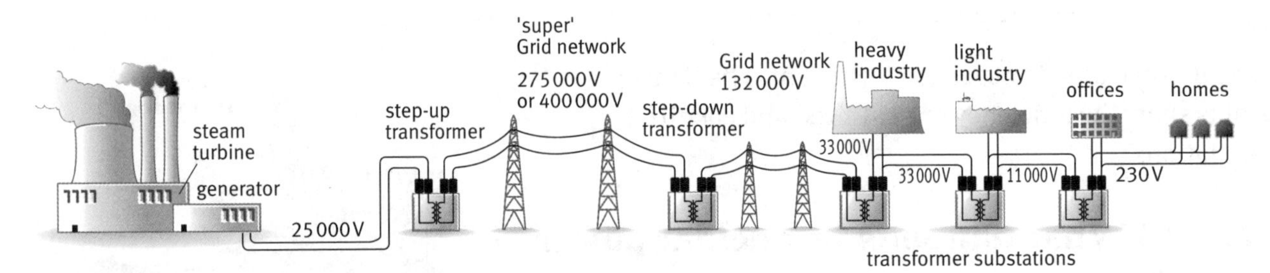

P3.1.4, P3.3.2–5 Which energy sources should we choose?

The choice of energy sources depends on factors including environmental impact, economics, waste, and CO_2 emissions.

H To be sure of a reliable electricity supply, a country needs a mix of energy sources.

When making decisions about generating electricity, energy planners need to consider:

- the expected power output of a power station
- the expected lifetime of a power station.

Exam tip

You might well be asked for advantages and disadvantages of different energy sources in the exam. Make sure you revise them carefully!

Use extra paper to answer these questions if you need to.

1 Tick the boxes to show which energy sources are renewable.

a coal ☐
b biofuels ☐
c waves ☐
d wind ☐
e geothermal ☐
f solar ☐

2 Draw lines to match each quantity to the correct units. Some quantities have more than one unit.

Quantity
voltage
energy
current
time
power

Unit
kilowatt-hour
joule
amp
watt
second
hour
volt

3 Put ticks in the boxes to show which energy sources each statement applies to.

	Biofuels	Gas	Solar
It is a primary energy source.			
When electricity is generated from it, CO_2 is produced.			
The generation of electricity from this source is weather dependent.			
When electricity is generated from it, no waste is produced.			

4 The statements below describe the energy flow in a nuclear power station. Write the letters of the steps in the correct order.

A Uranium atoms in the solid fuel split up.
B The hot fuel boils water.
C A magnet turns in a coil of wire.
D Steam drives turbines.
E An electric current flows.
F This releases energy and heats the fuel.
G This induces a voltage across the ends of the coil.

5 Calculate the missing numbers in the table. One kilowatt (kW) is 1000 watts (W).

Device	Power rating (W)	Power rating (kW)	Time it is on for	Energy transferred (kWh)
computer	250	0.250	2 h	
kettle	1800	1.800	3 min	
toaster			5 min	0.10
phone charger			2 h	0.04

6 List three ways in which an individual can reduce the amount of energy they use.

7 List three ways in which a government can reduce the amount of energy used in its country.

8 One unit (kWh) of electricity costs 10p. Calculate the cost of using the following electrical items:

a a 1.9 kW washing machine for 1.25 hours
b a 0.5 kW surround sound system for 2 hours.

9 Work out the missing numbers in the Sankey diagrams.

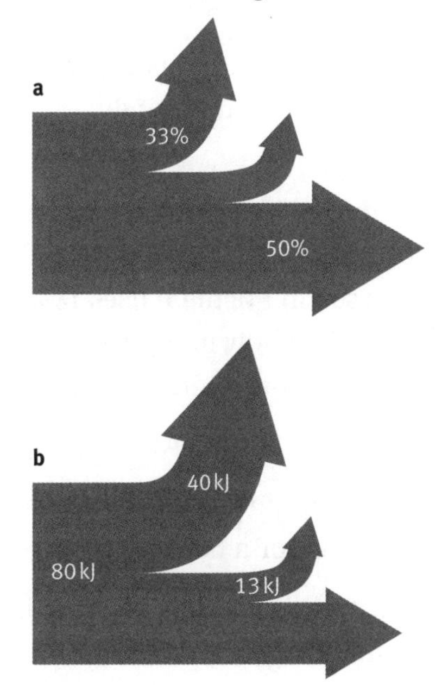

10 Explain why, in a Sankey diagram, the total width of the arrows must not change.

H **11** Write a definition for the **power** of an appliance or device. Include the word *rate* in your definition.

12 Explain why a country needs a mix of energy sources.

1 Read the article.

> **Portugal – leading the way in renewables**
> Between 2005 and 2010, the percentage of electricity generated from renewables in Portugal increased from 17% to 45%. This compares with 7% for the UK in 2010.
>
> In 2011, Portugal switched on mainland Europe's biggest wind farm. Portugal is soon to finish building the world's biggest solar power plant. The country is the first to generate electricity from wave power. Portugal also generates much electricity in hydroelectric power stations.

a Portugal's biggest wind farm has 120 turbines.
Each can supply 2 MW of electricity.

i Calculate the electrical power that can be supplied by all 120 turbines of the wind farm.

Answer = ______ MW [2]

ii Calculate the electrical energy supplied by the wind farm in 24 hours, if it is windy.

Answer = ______ MWh [2]

b The energy supplied by Portugal's biggest hydroelectric power station in one day is 15 120 000 kWh.
An average person in Portugal uses 13 kWh of electricity in one day. Calculate how many people the hydroelectric power station can supply with electricity for a day.

Answer = ______ people [2]

c Portugal's new solar power plant will have 2520 solar panels, each the size of a house. They are tilted.

i During one 24-hour period, each solar panel will turn around through an angle greater than 200°. Suggest why.

______________________________ [1]

ii The total energy supplied to a solar panel in one second is 120 000 J.
The useful energy transferred is 18 000 J.
Calculate the efficiency of a solar panel.

Answer = ______ % [2]

d The table gives the approximate efficiency for three methods of generating electricity.

Method of generating electricity	Efficiency
geothermal	16%
wind	59% of wind's energy extracted by blades
hydroelectric	90%

Suggest why Portugal uses wind power to generate electricity, as well as hydroelectric power stations, even though the efficiency for wind power is lower.

______________________________ [1]

e In 2008, Portugal opened the world's first wave farm. It is 5 km off the coast of Portugal. It converts energy in waves to electrical energy.

i Before it was built, many people supported the idea of having a wave farm. Suggest why.

______________________________ [1]

ii Some people were against the wave farm. Suggest why.

______________________________ [1]

Total [12]

2 A government wants to increase the amount of biofuels used to fuel cars, and decrease the amount of petrol and diesel.

Evaluate the benefits and problems of replacing fossil fuels with biofuels.

The quality of written communication will be assessed in your answer to this question.
Write your answer on separate paper or in your exercise book.

Total [6]

3 In 2008, a power station in Poland announced plans to expand.

The power station burns coal from its own mine, which is next to the power station.
The power station produces more carbon dioxide than any other power station in Europe.

a Use the words below to label the diagram of a coal-fired power station.
Write one word in each box.

furnace **steam** **generator** **transformer**

turbine **water**

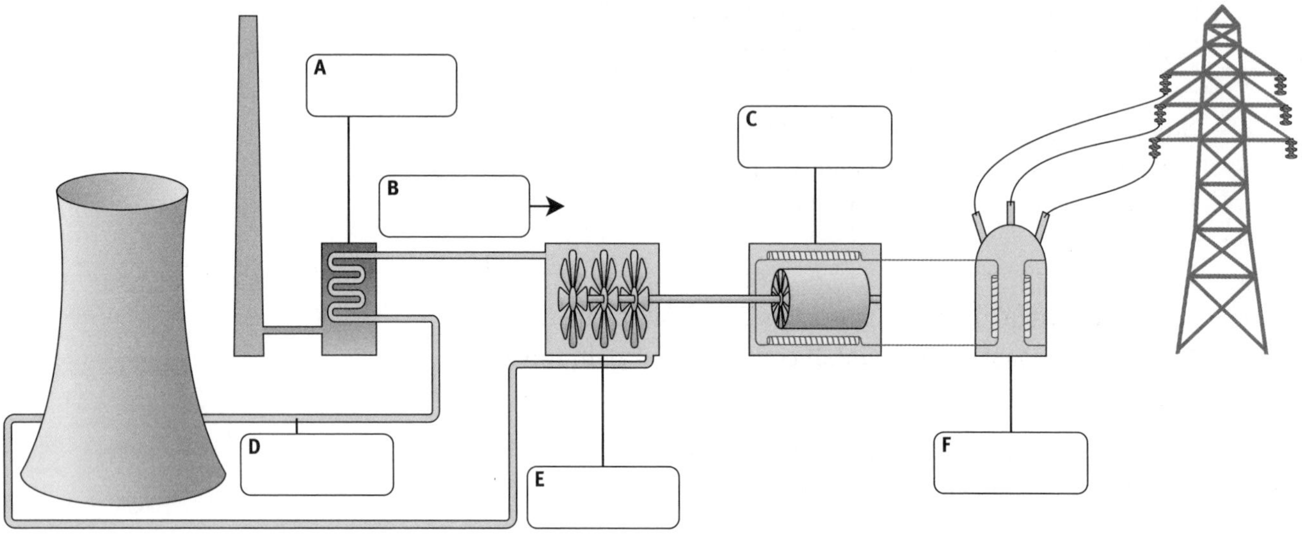

A

B

C

D

E

F

[3]

b The Sankey diagram below shows what happens to the energy in a coal-fired power station.

Calculate the percentage of energy that is carried away by exhaust gases.

Answer = ______ % [1]

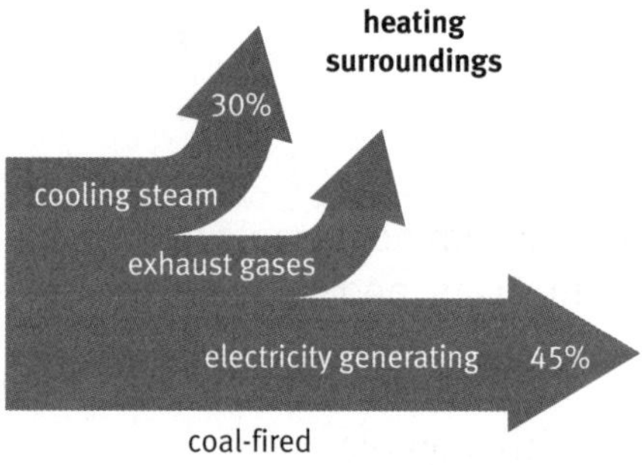

heating surroundings
30%
cooling steam
exhaust gases
electricity generating 45%
coal-fired

c Some people think the power station should not expand. Suggest one reason for this.

____

______________________________ [1]

d Some people think the power station should be replaced by a nuclear power station.
Give reasons for and against this idea.

____

____

____

____

______________________________ [3]

Total [8]

P4 Explaining motion

workout

1 David and Ruth are pushing on each other's hands. Neither person is moving.

Write **T** next the statements that are true.

Write **F** next to the statements that are false.

a The size of the force acting on David is less than the size of the force acting on Ruth. ☐

b The size of the force exerted by David is the same as the size of the force acting on Ruth. ☐

c Ruth experiences a bigger force than David. ☐

d Ruth and David exert forces of the same size. ☐

e The force exerted by Ruth is in the same direction as the force exerted by David. ☐

f The forces exerted by David and Ruth are opposite in direction. ☐

P4

2 Draw and label arrows to show the resultant forces on the rope, tricycle, and shopping trolley.

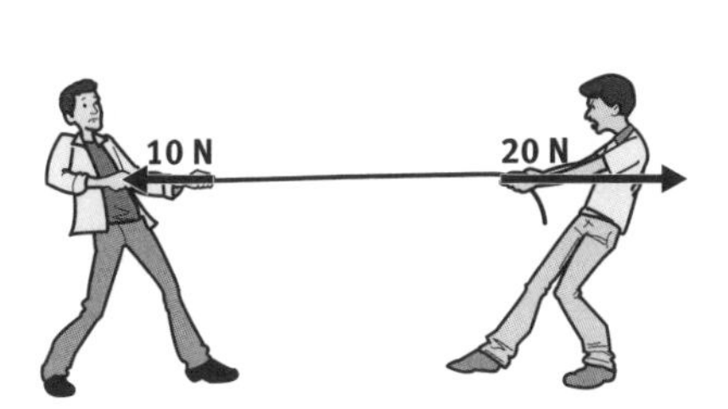

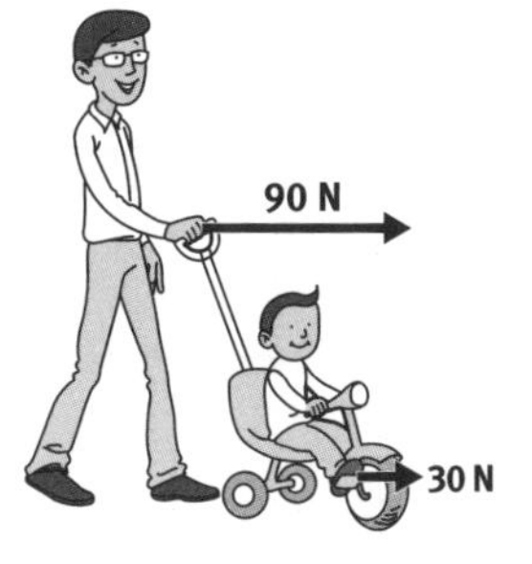

3 Faisal is moving a big loudspeaker.

Which caption belongs where? Write **A**, **B**, or **C** in each box.

A The friction force has reached its maximum.

B The size of the friction force is less than its maximum.

C There is no friction between the loudspeaker and the floor.

4 Kelly goes shopping at the mall. On the right is a distance–time graph for part of her time there.

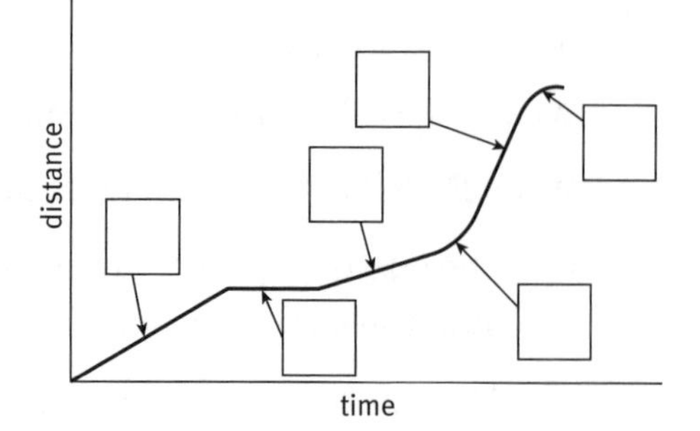

Label the graph by writing one letter in each box.

A standing still to look at shoes in a shop window

B walking quickly from the bus stop to the shops

C walking slowly past some clothes shops

D running at a constant speed

E starting to run when she realises she is late to meet her friend

F slowing down when she sees her friend in the distance

5 Solve the clues to fill in the grid.

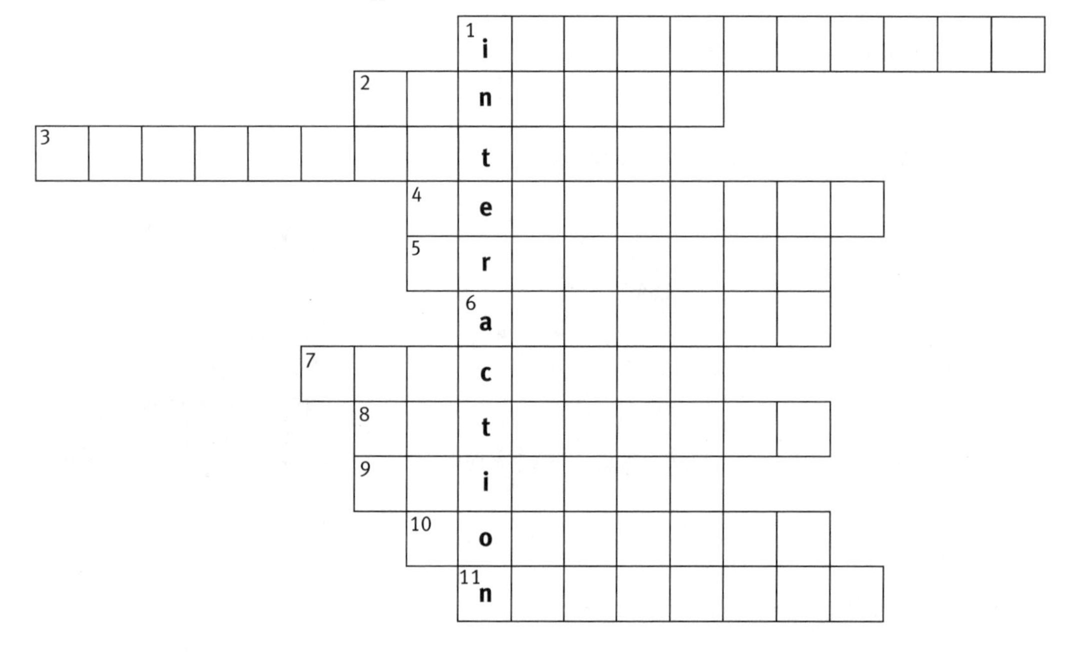

1 The two forces of an ______ pair are equal in size and opposite in direction.

2 A moving object has ______ energy.

3 The change in speed of an object in a given time interval is the ______ of the object.

4 Two people pull on a rope in opposite directions. The sum of the forces on the rope, taking direction into account, is the ______ force.

5 The force of ______ arises when you start pushing something over a surface.

6 Calculate the ______ speed of a car by dividing the total distance by the journey time.

7 A floor exerts a ______ force on a table leg that pushes down on it.

8 If you throw a basketball upwards, its gravitational ______ energy increases.

9 The force that makes you move forwards on a scooter is the ______ force.

10 Multiplying the mass of a train by its velocity gives you the train's ______.

11 If a football travels in a straight line in one direction, its velocity is positive. When it moves in the opposite direction, its velocity is ______.

What are forces?

P4.2.1–4 Interaction pairs

Lucy and Luke push against each other. They are not moving. Lucy exerts a force on Luke. Luke exerts a force on Lucy.

- Forces arise from an **interaction** between two objects. They come in pairs.
- Each force in an **interaction pair** acts on a different object. The forces are **equal** in size and **opposite** in direction.

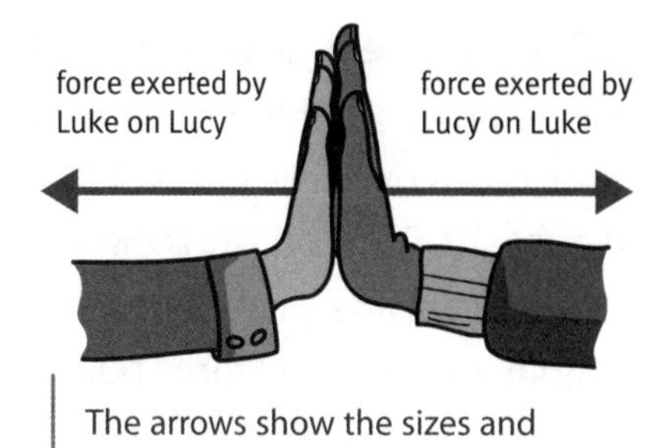

The arrows show the sizes and directions of the forces.

P4.3.1–2 Resultant force

The **resultant force** on an object is the sum of the individual forces that act on it, taking their directions into account.

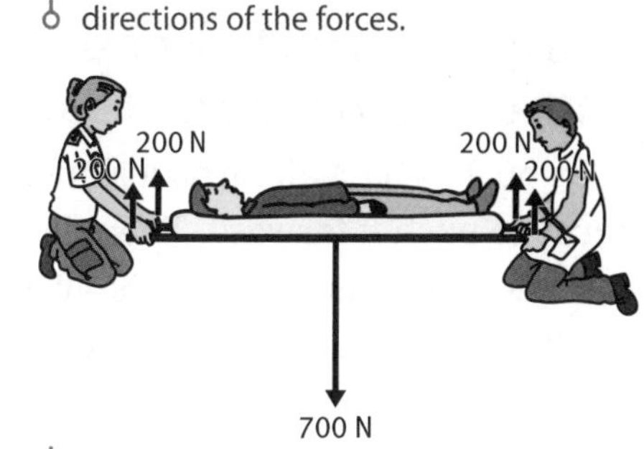

The resultant force on the stretcher is 100 N in an upward direction. The stretcher starts moving up from the ground.

P4.2.6 Reaction of surfaces

Vincent's feet push down on the floor. The floor pushes up on his feet with an equal force. This force is the **reaction of the surface**.

P4.2.5, P4.2.7 Friction

David tries to push a skip. The force of **friction** stops the skip sliding over the road's surface.

As David pushes harder, the size of the friction force increases. Eventually, the friction force reaches its limit. Now the skip moves.

There was no friction force between the skip and the road before David tried to push the skip. Friction arose in response to the force that David applied.

The friction force balances David's push. The skip does not move.

The friction force balances David's push. The skip still does not move.

The skip moves. 6000 N is bigger than the maximum possible friction force for this skip and the road surface.

How do objects start moving?

P4.2.8 Using friction

When a car engine starts, the wheels turn. They exert a backward-pushing force on the road surface. The other force in the interaction pair, the forward force, is the same size. This gets the car moving.

When you walk, your foot pushes back on the ground. The friction between your foot and the ground pushes you forward with an equal force.

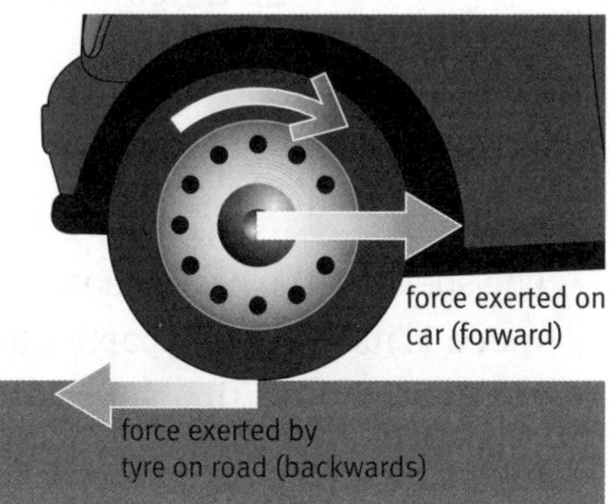

The tyre grips the road. The road exerts a large forward force on the axle. This pushes the car forward.

P4.2.9 Rockets and jet engines

A rocket pushes out hot gases as its fuel burns. The rocket pushes down on these gases. The escaping gases exert an equal and opposite force on the rocket, and push the rocket upwards.

A jet engine draws in air at the front and pushes it out at the back. An equal and opposite force pushes the engine forwards.

Why do objects keep moving?

P4.3.6–7 Driving and counter forces

Alex pushes Sam along on a skateboard. Alex exerts the **driving force** to push it forward. There is a **counter force** in the opposite direction, because of air resistance and friction.

- If the driving force is greater than the counter force, the skateboard speeds up.
- If the driving force is equal to the counter force, the skateboard moves at a constant speed in a straight line.
- If the driving force is less than the counter force, the skateboard slows down.

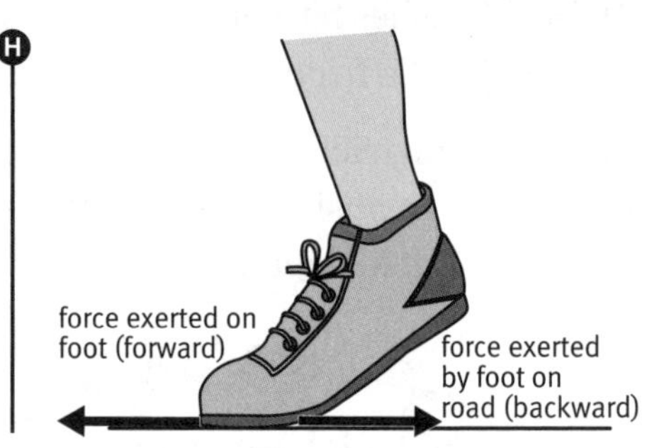

P4.1.1–2, P4.1.9 Speed and velocity

You can use the equation below to calculate **average speed**.

$$\textbf{speed}\ (\text{m/s}) = \frac{\textbf{distance travelled}\ (\text{m})}{\textbf{time}\ (\text{s})}$$

So if a horse runs 20 metres in 10 seconds:

$$\text{average speed} = \frac{20\ \text{m}}{10\ \text{s}} = 2\ \text{m/s}$$

The speed of the horse changes as it runs. Its **instantaneous speed** is its speed at a particular instant, or its average speed over a very short time interval.

The **instantaneous velocity** of an object is its instantaneous speed in a certain direction.

P4.1.8, P4.1.13 Acceleration

If a car gets faster, it is accelerating. The **acceleration** of the car is its change of speed, or change of velocity, in a given time interval. You can use the equation here to calculate acceleration:

$$\textbf{acceleration}\ (\text{m/s}) = \frac{\textbf{change in velocity}\ (\text{m/s})}{\textbf{time taken}\ (\text{s})}$$

If a car accelerates from 10 m/s to 30 m/s in 10 s:

$$\text{acceleration} = \frac{(30 - 10)\ \text{m/s}}{10\ \text{s}}$$
$$= 2\ \text{m/s}^2$$

How can we describe motion?

4.1.3–6 Distance–time and displacement–time graphs

Distance–time graphs describe movement.

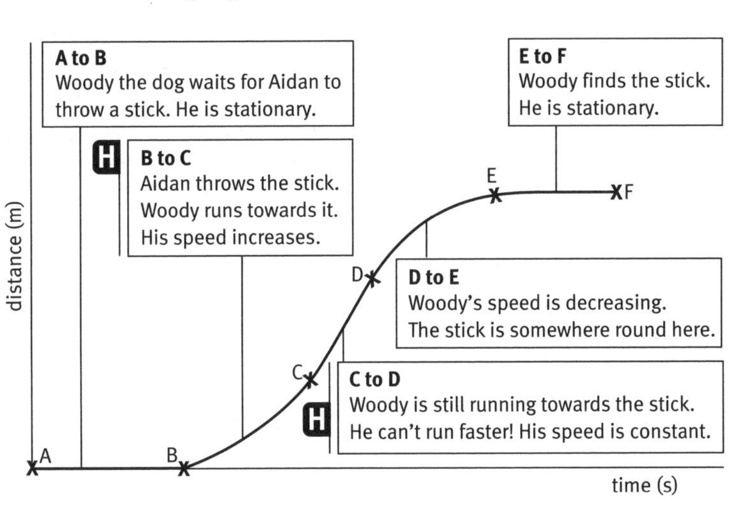

A to B
Woody the dog waits for Aidan to throw a stick. He is stationary.
B to C
Aidan throws the stick. Woody runs towards it. His speed increases.
C to D
Woody is still running towards the stick. He can't run faster! His speed is constant.
D to E
Woody's speed is decreasing. The stick is somewhere round here.
E to F
Woody finds the stick. He is stationary.
distance (m)
time (s)
A
B
C
D
E
F

You can use distance–time graphs to calculate speed.

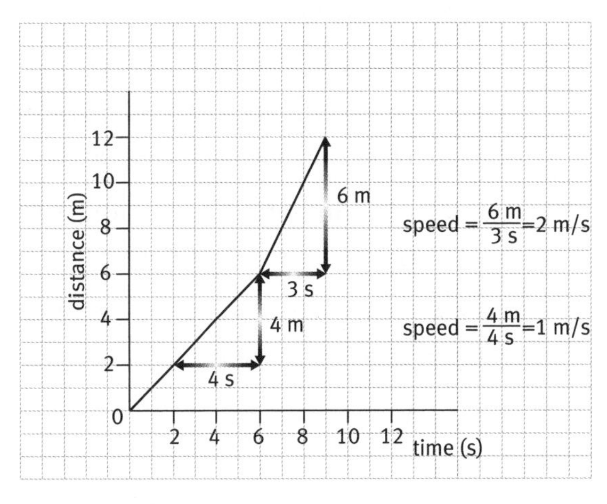

The steeper the gradient of a distance–time graph, the greater the speed.

The **displacement** of an object at a given moment is the straight-line distance from its starting point, with an indication of direction. Freya throws a ball into the air. The graphs show how its distance and displacement change with time.

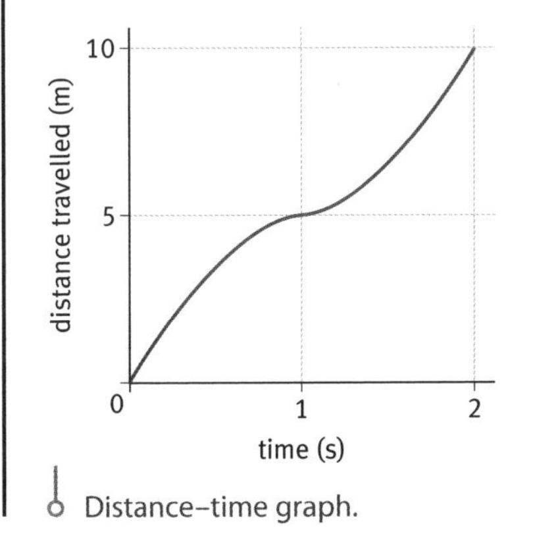

Distance–time graph.

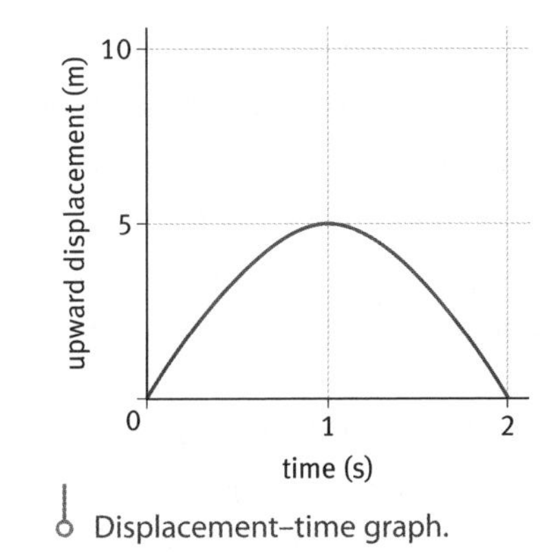

Displacement–time graph.

P4.1.7, P4.1.10–12 Speed–time and velocity–time graphs

Speed–time graphs show how speed varies with time. The graph opposite shows the speed of a train on a 3-hour journey.

Velocity–time graphs show the velocity of an object at every instant of its journey. The graph below shows the velocity of Ella, an ice dancer. The gradient of a section of the graph is equal to Ella's acceleration.

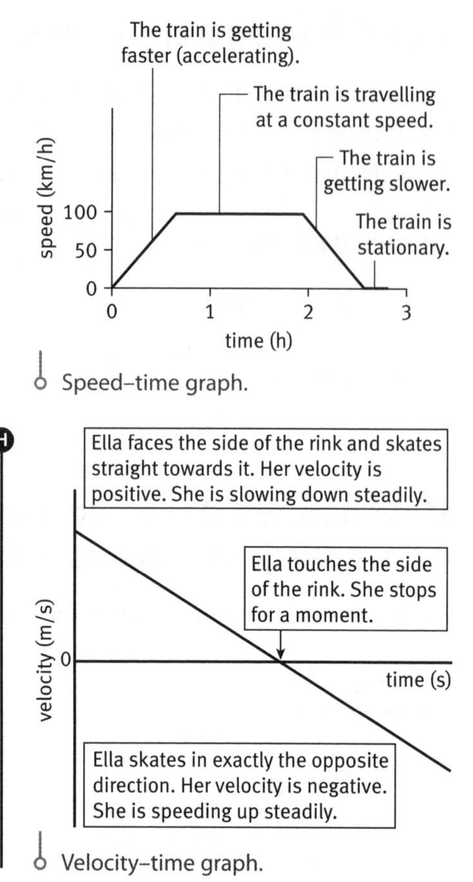

The train is getting faster (accelerating).
The train is travelling at a constant speed.
The train is getting slower.
The train is stationary.
speed (km/h)
100
50
0
0
1
2
3
time (h)
Ella faces the side of the rink and skates straight towards it. Her velocity is positive. She is slowing down steadily.
Ella touches the side of the rink. She stops for a moment.
velocity (m/s)
0
time (s)
Ella skates in exactly the opposite direction. Her velocity is negative. She is speeding up steadily.

Speed–time graph.

Velocity–time graph.

How are forces and motion connected?

P4.3.3–5, P4.3.11 Momentum

All moving objects have **momentum**.

momentum (kg m/s) = **mass** (kg) × **velocity** (m/s)

For a 0.5 kg bird flying at a velocity of 2 m/s:

momentum = 0.5 kg × 2 m/s
= 1 kg m/s

When a resultant force acts on an object, the momentum of the object changes in the direction of the force:

change of momentum = resultant force × time for which it acts
(kg m/s) (newton, N) (second, s)

If a 3-second gust of wind from behind a bird exerts a resultant force of 10 N on the bird:

change of momentum = 10 N × 3 s
= 30 kg m/s in the direction the bird is flying

If the resultant force on an object is zero, its momentum does not change:

- If it is stationary, it stays still.
- If it is already moving, it continues at a steady speed in a straight line.

P4.3.8–9 Road safety

If two cars collide and stop, their momentum changes until it becomes zero. The more time the change of momentum takes, the smaller the resultant force. Road safety measures use this idea:

- Car **crumple zones** squash slowly in a collision. So the collision lasts longer, and the resultant force on the car is less.
- **Seat belts** stretch in a collision. This makes the change of momentum take longer. So the force is less.
- **Helmets** change shape when they hit something. Your head stops moving more slowly, so the force on it is less.
- **Air bags** also increase the time for the change of momentum.

How can we use energy changes to describe motion?

P4.4.1, P4.4.3–5, P4.4.8–13 Doing work

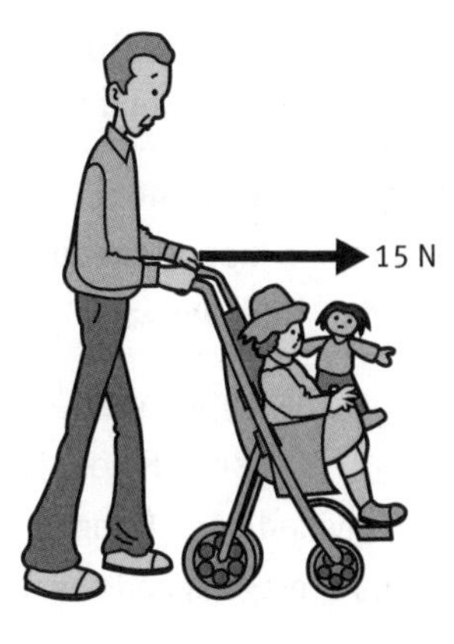

Barney takes his child to the park. He pushes the buggy with a force of 15 N. The force makes the buggy move. Barney is **doing work**.

work done by a force = force × distance moved by the force
(joule, J) (newton, N) (metre, m)

The park is 500 m away from Barney's house. So:
work done by Barney = 15 N × 500 m = 7500 J

Barney **transfers energy** to the buggy. His store of chemical energy decreases. The energy of the buggy increases.

amount of energy transferred = work done
(joule, J) (joule, J)

amount of energy transferred by Barney = work done = 7500 J

The moving buggy has **kinetic energy (KE)**. Kinetic energy depends on **mass** and **velocity**.

kinetic energy = ½ × mass × (velocity)²
(joule, J) (kilogram, kg) (metre per second, m/s)²

So the faster the buggy moves, and the greater the mass of the child in it, the more kinetic energy it has.

If Barney pushes with a greater force, he does more work and so transfers more energy. The buggy goes faster and its kinetic energy increases.

In fact, the gain of kinetic energy by the buggy is less than the energy transferred from Barney. Barney must also transfer enough energy to overcome air resistance and friction. This energy is transferred to the surroundings as heat.

Overall, as in every event and process, energy is conserved.

P4.3.10, P4.4.2,6,7,14 Gravitational potential energy

Catherine picks up her doll from the ground. She is doing work. The doll's **gravitational potential energy (GPE)** increases.

change in GPE = weight × vertical height difference
(joule, J) (newton, N) (metre, m)

The doll's weight is 3 N. Catherine lifts it 1 m. So:

change in doll's GPE = 3 N × 1 m = 3 J

Catherine lets go of the doll. The force of gravity pulls it downwards. The doll falls 1 m to the ground. Its kinetic energy increases and its GPE decreases.

GPE lost = kinetic energy gained

So the doll gains 3 J of kinetic energy.

H To calculate the doll's speed as it falls:

KE = ½ × mass × (velocity)²

The doll's mass is 0.3 kg.
Rearranging gives:

$$\text{velocity} = \sqrt{\frac{2 \times \text{kinetic energy}}{\text{mass}}}$$

$$= \sqrt{\frac{(2 \times 3\text{ J})}{0.3\text{ kg}}}$$

$$= \sqrt{(20\text{ J/kg}}$$

$$= 4.5\text{ m/s}$$

1 Use the words in the box to fill in the gaps. Each word may be used once, more than once, or not at all.

average	**constant**	**increasing**	**gravity**
friction	**instantaneous**	**short**	**long**
upwards	**downwards**	**decreasing**	**time**

A bird of prey flies from one tree to another in a straight line. Its _______ speed is equal to the distance between the trees, divided by its flying _______. Its instantaneous speed is its average speed over a very _______ time interval.

The bird drops to the ground. The force of _______ pulls the bird downwards. Air resistance exerts an _______ force on the bird.

The bird takes off again and flies along at a steady speed in a straight line. While it flies along, its momentum is _______.

2 Saima pulls along her suitcase.
The arrows show the directions of the counter force and the driving force.

Write **T** next to the statements that are true.
Write **F** next to the statements that are false.

- **a** If the driving force is less than the counter force, the suitcase slows down.
- **b** Saima exerts a driving force to pull along the suitcase.
- **c** If the driving force is equal to the counter force, the suitcase moves with a constant speed.
- **d** The counter force is caused by air resistance only.
- **e** If the driving force is more than the counter force, the suitcase speeds up.
- **f** The counter force is caused by air resistance and friction.
- **g** If the driving force is equal to the counter force, the suitcase cannot move.

3 Highlight the correct word in each pair of **bold** words.

- **a** The faster an object moves, the **smaller / greater** its kinetic energy.
- **b** Ali lifts a weight. Energy is transferred from **the weight / Ali** to **the weight /Ali**.
- **c** Harry pulls a toy sledge across the carpet. Friction causes the sledge to gain **less / more** kinetic energy than the work Harry put in to pulling the sledge, because some energy is dissipated through **heating / pulling**.

4 Calculate the average speed of the following. Include units in your answers.

- **a** a helicopter that travels 600 metres in 3 minutes
- **b** a football that travels 80 metres in 2 seconds
- **c** a racehorse that runs 900 metres in 50 seconds
- **d** a worm that moves 32 centimetres in 8 seconds

5 Calculate the acceleration of the following. Include units in your answers.

- **a** a runner whose velocity increases from 0 m/s to 10 m/s in 2 seconds
- **b** a car whose velocity increases from 10 m/s to 30 m/s in 10 seconds

6 Calculate the momentum of each of the following:

- **a** a 2000 kg sports car moving at a velocity of 44 m/s
- **b** a 70 kg person on a 6 kg scooter moving at a velocity of 4 m/s
- **c** a 9 kg baby crawling at a velocity of 1.5 m/s

7 A driver does an emergency stop as a child runs out in front of her car. The car stops in 3 seconds. The resultant force on the car is 5000 N. Calculate the change in momentum.

8 Calculate the kinetic energy of each of the following:

- **a** a 150 kg lion running with a velocity of 20 m/s
- **b** a 4000 kg bus moving with a velocity of 25 m/s
- **c** a 60 g tennis ball moving with a velocity of 44 m/s

9 Use the idea of a pair of equal and opposite forces to explain how jet engines produce a driving force.

H 10 Pawel is a swimmer.
Here is his velocity–time graph for the first 55 m of a 100-m race.

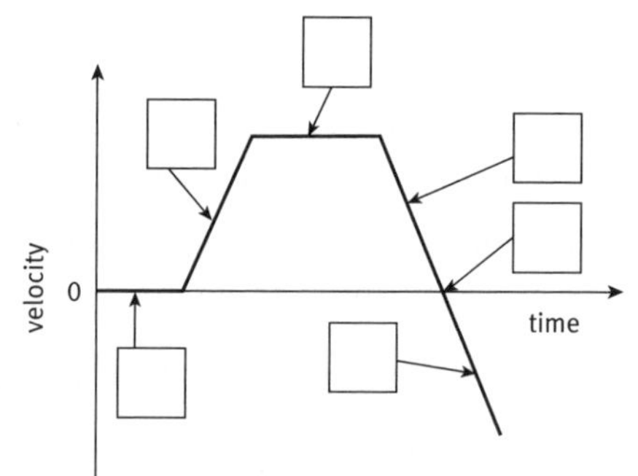

Label the graph by writing one letter in each box.

- **A** Pawel is waiting to dive in. He is stationary.
- **B** Pawel is moving in a straight line. His speed is steadily increasing.
- **C** Pawel has turned round. He has changed direction. His speed is steadily increasing.
- **D** Pawel is moving in a straight line. He is swimming at a constant speed.
- **E** Pawel is turning round. He is stationary for an instant.
- **F** Pawel is moving in a straight line. He is slowing down.

1 A penguin stands at the top of a slope. It slides to the bottom.

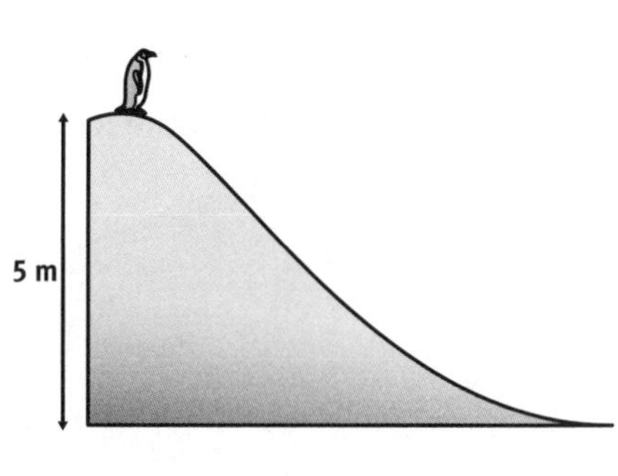

a **i** As the penguin slides down the slope its gravitational potential energy (GPE) and its kinetic energy (KE) change.

Tick the statements that are true.

The GPE of the penguin at the top of the slope is less than its GPE at the bottom. ☐

As the penguin slides down the slope, its GPE decreases. ☐

The force of gravity does work on the penguin as it slides down the slope. ☐

The penguin's velocity increases as it slides down the slope. ☐

As the penguin slides down the slope, it gains KE. ☐ [2]

ii Calculate the change in the penguin's gravitational potential energy.

The weight of the penguin is 300 N.

Change in GPE = ________________ J [2]

iii Assume that friction is small enough to ignore.

What is the change in the penguin's kinetic energy?

Change in KE = ________________ J [1]

b A baby penguin travels down the same slope.

Its mass is 6 kg.

Its velocity at the bottom of the slope is 10 m/s.

Calculate the kinetic energy of the baby penguin at the bottom of the slope.

KE = ________________ J [2]

Total [7]

Exam tip

Don't forget to show your working in calculations. If your answer is wrong, but your working is correct, you might still get a mark.

2 Look at the pictures of the lorries.

Lorry A and lorry B have the same mass and their tyres are the same.

The shape of lorry A means that it experiences less air resistance than lorry B.

Use ideas about forces in interaction pairs to explain why the driving force needed by lorry B is less than the driving force needed by lorry A, when they travel in a straight line at a steady speed.

Assume the loads on the two lorries, and the weather and road conditions, are the same for both lorries.

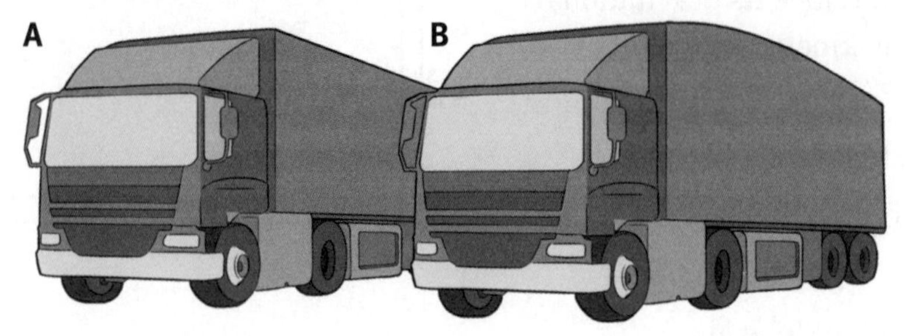

The quality of written communication will be assessed in your answer to this question.
Write your answer on separate paper or in your exercise book.

Total [6]

Exam tip

When answering 6-mark questions:
- Make sure you know what the question is asking you to do.
- Write down key words to help answer the question.
- Organise the key words in a sensible order.
- Write your answer.
- Check your spelling, punctuation, and grammar.

3 A fire engine travels to a fire.

The graph shows its journey.

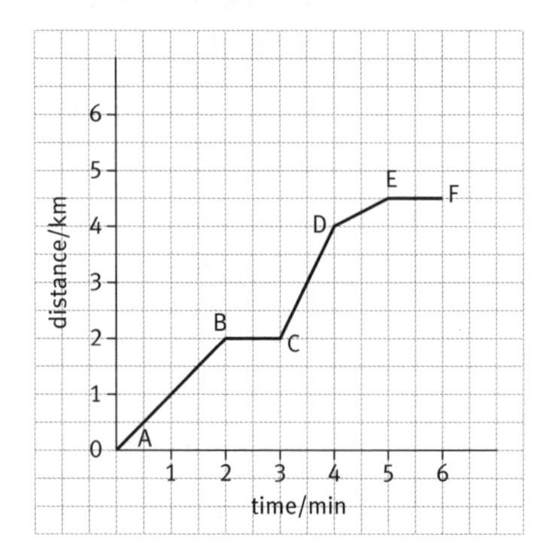

a i In which part of the journey was the fire engine moving along most slowly?

Draw a (ring) around the correct answer.

A to B **B to C** **C to D** **D to E** **E to F** [1]

ii Describe the motion of the fire engine from **B to D**.

____

______________________________ [2]

iii Calculate the average speed of the fire engine between **A and B**.

Average speed = ______________ km/minute [2]

b A police car travels to the same fire. It goes 6000 metres in 500 seconds.

Calculate the average speed of the police car.

Average speed = ______________ m/s [2]

Total [7]

4 Oona used a book to find the fastest speeds of some animals. She wrote the speeds in a table.

Animal	Fastest speed (m/s)
cheetah	28
horse	17
cat	13
elephant	11
pig	6
chicken	4
mouse	3
snail	0.01

Opposite are distance–time and speed–time graphs for some of the animals in the table.

On the graphs, the animals in the table above are represented by letters.

a On the **speed–time graph**, animal E is a chicken and animal H is a cheetah.

What might be the identities of animals F and G?

Put a tick (✓) in the box next to the **one** correct statement.

F = snail G = chicken ☐

F = cat G = horse ☐

F = cat G = chicken ☐

F = chicken G = snail ☐ [1]

b Use the **distance–time graph** to calculate the top speeds of animals A and B.

____

____________________ [2]

c Which of the following are correct conclusions from the distance–time graph?

Put ticks (✓) in the boxes next to the **two** correct statements.

The top speed of animal B is twice the top speed of animal A. ☐

There is no animal in the world that has a slower top speed than animal A. ☐

The top speed of animal D is 11 m/s. ☐

Animal C is unlikely to be a pig. ☐

The graph shows that animal D is thefastest animal in the world. ☐ [2]

Total [5]

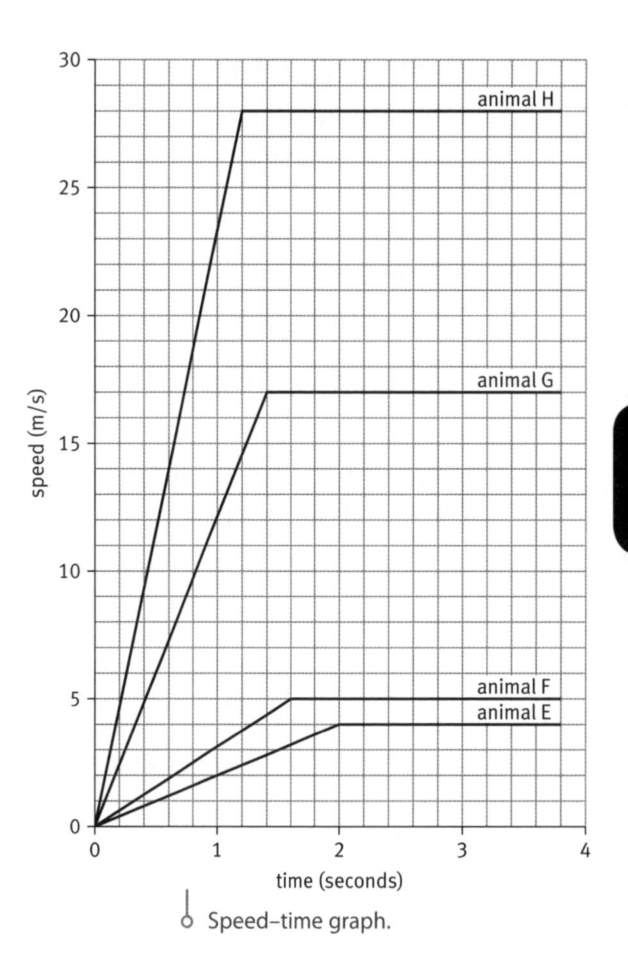

Speed–time graph.

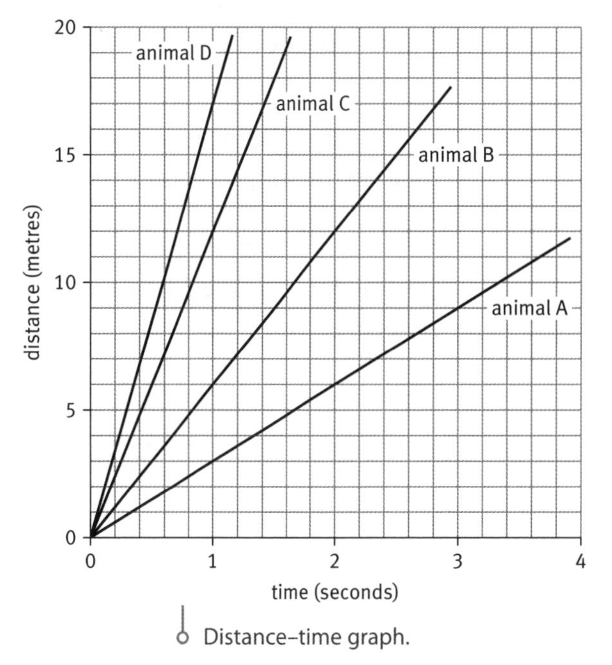

Distance–time graph.

P 4

Going for the highest grades

H 5 A motorbike goes along a straight test track.

- For the first 6 seconds it accelerates until it reaches a speed of 40 m/s.
- For the next 4 seconds it moves at a steady speed.
- Then it slows down.
- It stops 14 seconds after its journey began.

a Finish the velocity–time graph for the journey: [3]

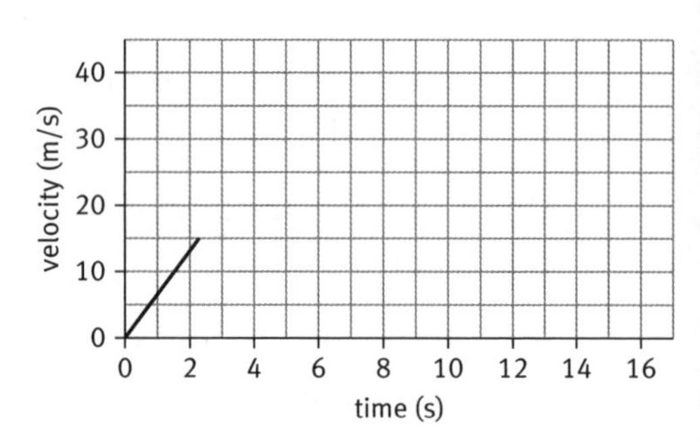

b Calculate the acceleration of the motorbike during the first 6 seconds of its journey.

Answer = ______________ m/s^2 [2]

c Another motorbike travels along the same straight test track at a speed of 30 m/s. It then turns round and goes back along the same test track at the same speed.

Describe the change in velocity of the motorbike.

____

____ [1]

d A third motorbike travels along the test track.

The graph opposite shows how the displacement of the bike changes with time.

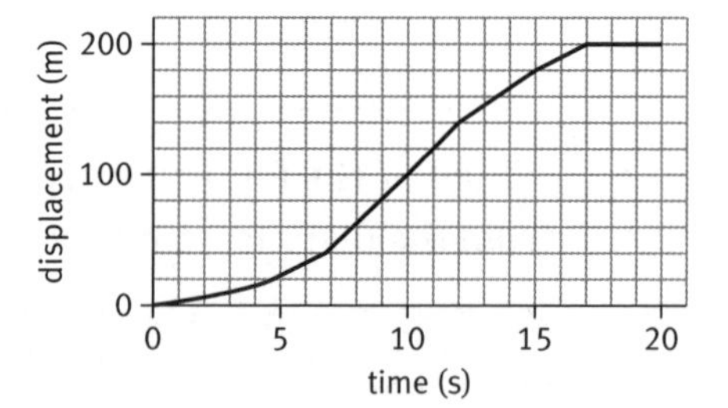

i What is the displacement of the bike at 10 seconds? Include the unit in your answer.

____ [1]

ii Describe the motion of the motorbike between 12 and 15 seconds.

____

____ [2]

Total [9]

P5 Electric circuits workout

1 Fill in the gaps.

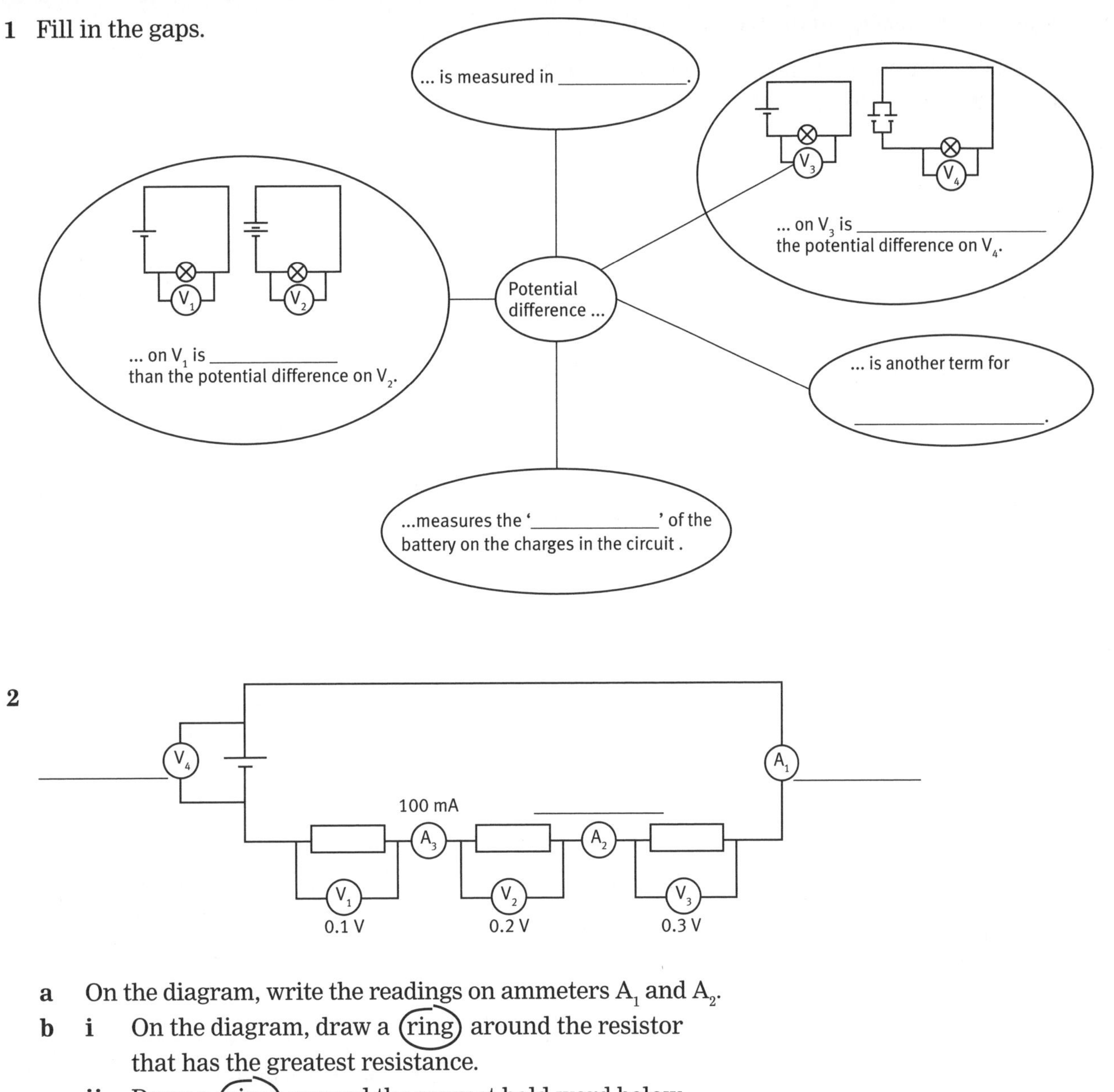

a On the diagram, write the readings on ammeters A_1 and A_2.

b **i** On the diagram, draw a (ring) around the resistor that has the greatest resistance.

ii Draw a (ring) around the correct bold word below. Then complete the sentence.

The potential difference is **smallest / greatest** across the component with the greatest resistance because

______

c **i** On the diagram, write the reading on voltmeter V_4.

ii Complete the sentence:

I know that this is the voltage on V_4 because ______

______

______

Exam tip

Before you answer a question about a circuit, check whether the components are connected in series or in parallel.

P 5

3 In the six circuits below, all the lamps are identical.

For each pair of circuits, draw a (ring) around the circuit in which the ammeter reading is greater.

a

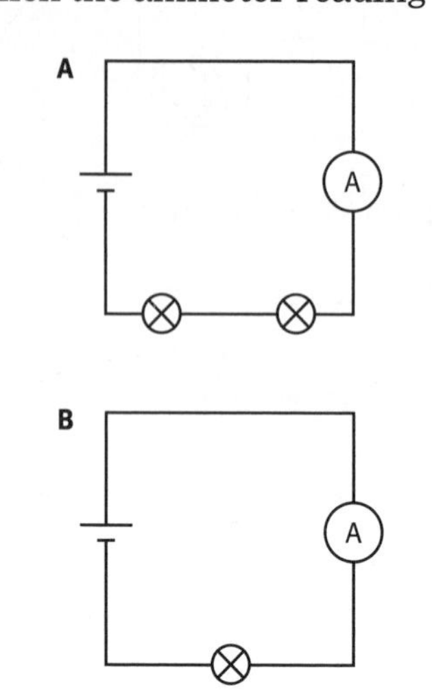

b

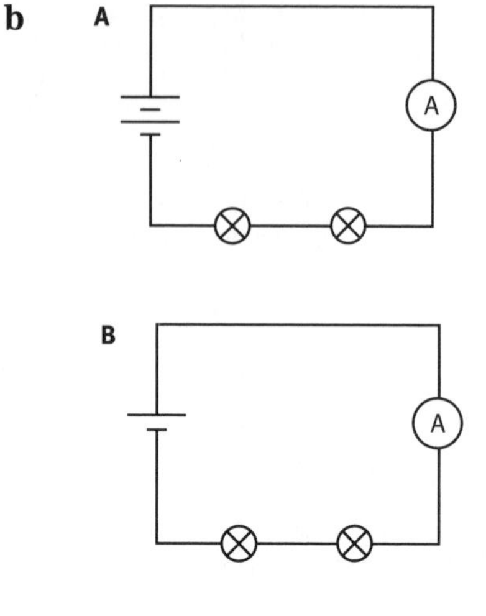

c

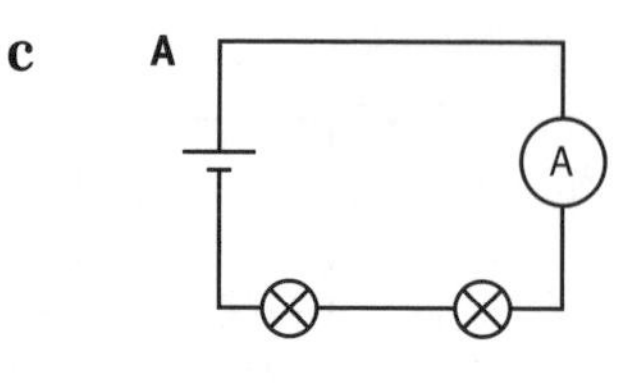

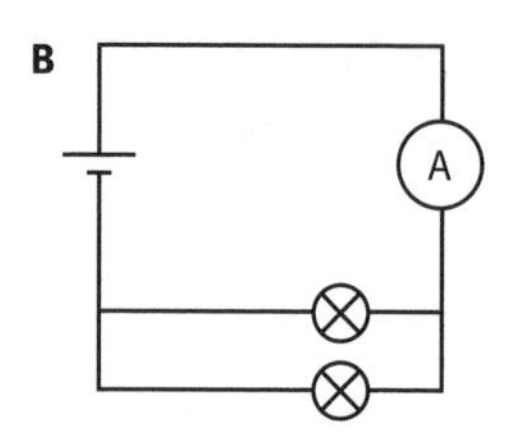

4 Solve the clues to fill in the arrow words.

1 →								2 ↓	
3 →		4 →							5 →
						← 6			16 ↓
7 →									
8 →									← 9
10 →									11 →
12 →						13 →			14 →
15 →									

Horizontal

1 Divide the voltage by the current to calculate this.

3 The unit of resistance.

4 The rate at which a power supply transfers energy to an appliance.

5 The symbol for the unit of potential difference.

6 In this type of circuit, the current through each component is the same as if it were the only component.

7 Use this device to measure potential difference across a component in a circuit.

8 The symbol for the unit of resistance.

9 This device consists of a magnet rotating within a coil of wire.

10 Generators produce electricity by electromagnetic ______

11 The symbol for resistance.

12 Batteries produce ______ current.

13 This type of current reverses direction several times a second.

14 The symbol for the unit of electric current.

15 In a motor, a ______ reverses the direction of the current in the coil at an appropriate point in each revolution.

Vertical

2 A flow of charge.

16 The abbreviation for direct current is ______.

P5.1.1–4 What is static electricity?

If you rub a balloon in your hair, the balloon and your hair become charged. Tiny negative particles (electrons) move from your hair to the balloon.

Each hair is positively charged. Like charges repel. So the hairs get as far away from each other as possible.

There are attractive forces between opposite charges, so positively charged hairs are attracted to the negatively charged balloon.

P5.1.5–10 What is an electric current?

Electric current is a flow of charge.

Metal conductors have many charges (electrons) that are free to move. Electric current is the movement of these free electrons. Insulators do not conduct electricity. This is because they have very few charges that are free to move.

The components and wires conduct electricity when the switch is closed and the circuit is complete.

The battery makes free charges move in a continuous loop. The free charges are not used up.

The ammeter measures electric current in amperes, or amps (A).

P5.2.5–7 What is electrical power?

When an electric charge flows through a component or device in a circuit, work is done by the power supply. Energy is transferred from the power supply to the component and its surroundings.

Power is a measure of the rate at which an electrical power supply transfers energy to an appliance or device, and its surroundings.

You can use the equation below to calculate power:

power	= **voltage**	× **current**
(watts, W)	(volts, V)	(amperes, A)

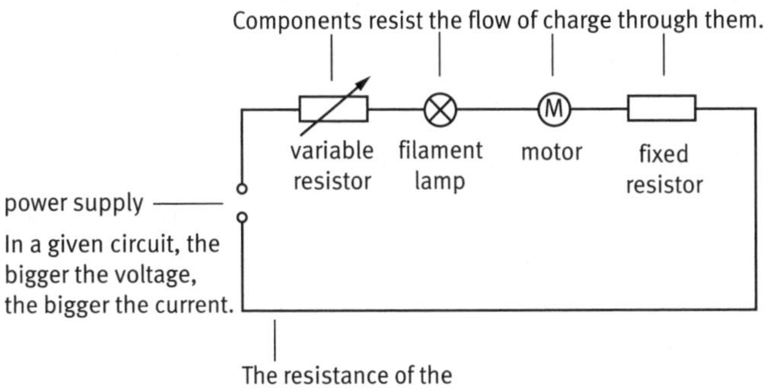

Components resist the flow of charge through them.
variable resistor
filament lamp
motor
fixed resistor
power supply
In a given circuit, the bigger the voltage, the bigger the current.
The resistance of the connecting wires is tiny, so you can usually ignore it.

P5.2.1–4, P5.2.8–16 Resistance?

The bigger the resistance, the smaller the current.

The current through a metal conductor is proportional to the voltage across it:

$$\textbf{resistance}\ (\text{ohm}, \Omega) = \frac{\textbf{voltage}\ (\text{volt, V})}{\textbf{current}\ (\text{ampere, A})}$$

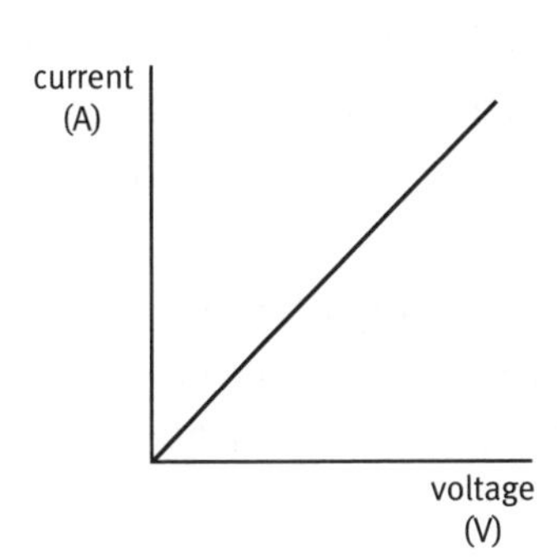

The gradient of the graph is constant. It is equal to the resistance of the resistor.

Resistors get hotter when electric current passes through them. This happens because moving electrons bump into stationary ions in the wire.

Lamp filaments get so hot that they glow.

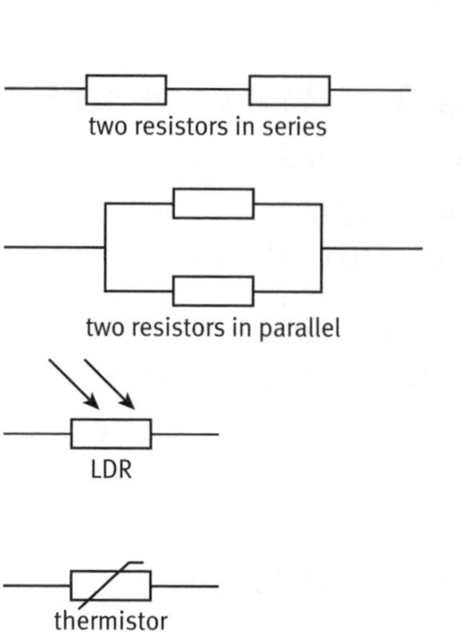

Two resistors in **series** have more resistance than one on its own. The battery must push charges through both resistors.

Two resistors in **parallel** have a smaller total resistance than one on its own. There are more paths for electric charges to flow along.

The resistance of a light-dependent resistor (**LDR**) changes with light intensity. Its resistance in the dark is greater than its resistance in the light. LDRs switch outdoor lights on at night.

The resistance of a **thermistor** changes with temperature. For many thermistors, the hotter the temperature, the lower the resistance. Thermistors switch water heaters on and off.

5.3.1–7 How do series and parallel circuits work?

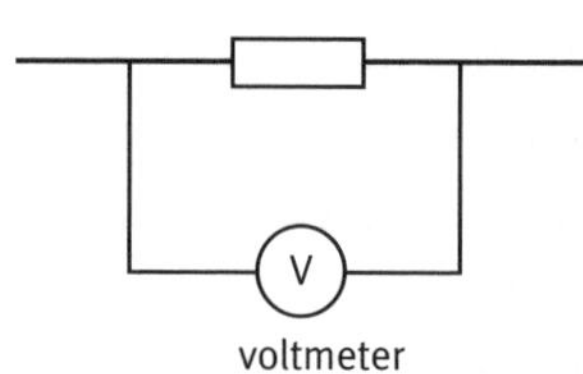

The **voltage** of a battery shows its 'push' on the charges in a circuit. **Potential difference (p.d.)** means the same as voltage.

The greater the potential difference between two points in a circuit, the more work must be done to make a given amount of charge move between the points.

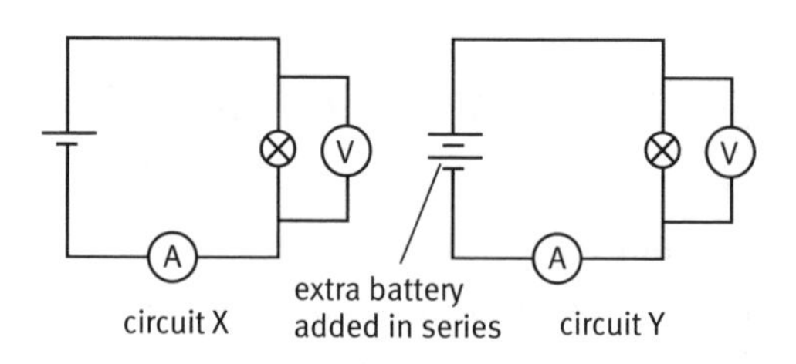

You can use a **voltmeter** to measure the potential difference across a component in a circuit. The diagram (above right) shows how to connect it.

The readings on the voltmeter and ammeter are greater in circuit Y than in circuit X (see right). The second battery gives an extra 'push' to the charges in the circuit.

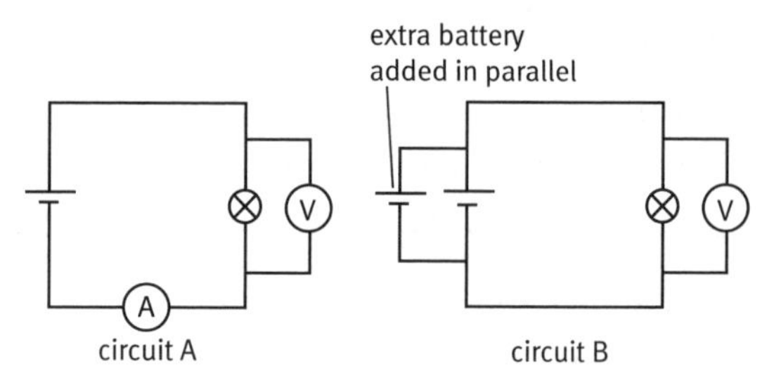

The readings on the voltmeter and ammeter are the same in circuits A and B (see right). The second battery provides no extra 'push' to the charges in the circuit.

In circuit S:

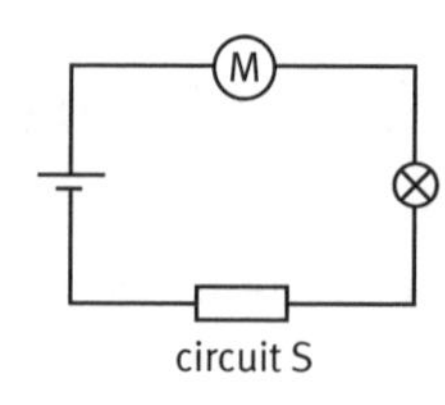

- Three components are connected in series to a battery.
- The same current flows through each component.
- The potential differences across the components add up to the potential difference across the battery.
  This is because the work done on each unit of charge by the battery must equal the work done by it on the circuit components.
- The potential difference is biggest across the component with the greatest resistance.
- This is because a charge flowing through a big resistance does more work than a charge flowing through a smaller resistance.
- If you change the resistance of one of the components, there will be a change in the potential differences across all the components.

In circuit P:

- Three components are connected in parallel to a battery.
- The current at J, and at K, is equal to the sum of the currents through the components.
- The current is largest through the component with the smallest resistance.

  This is because the same battery voltage pushes more current through a component with a smaller resistance than through one with a bigger resistance.
- The current through each component is the same as if it were the only component.
- The p.d. across each component is the same as p.d. of the battery.

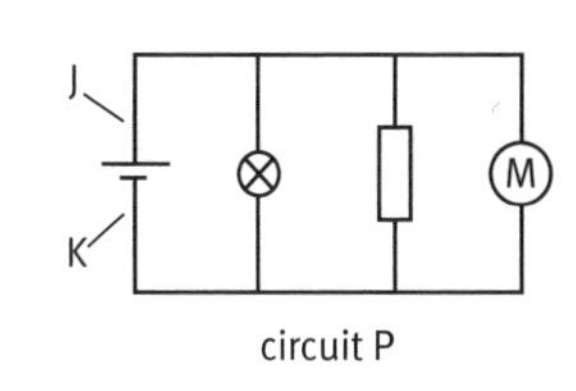

circuit P

P5.4.3–5 What is electromagnetic induction?

If you move a magnet into a coil of wire, a voltage is induced across the ends of the coil (diagram A). This is **electromagnetic induction**. If you join up the ends of the coil to make a circuit, a current flows.

You can induce a voltage in the opposite direction by:

- moving the magnet *out* of the coil (diagram B) or
- moving the *other pole of* the magnet into the coil (diagram C).

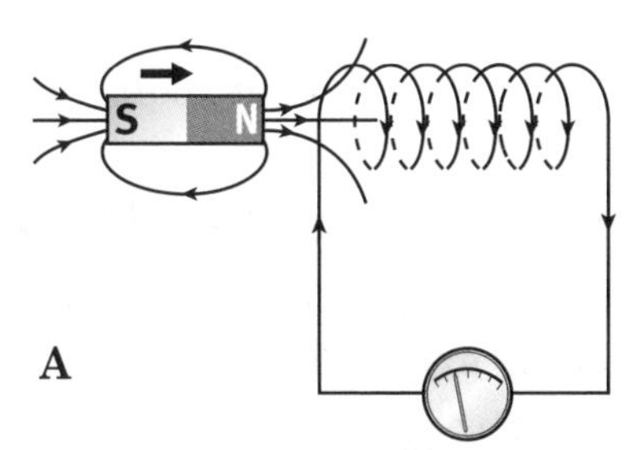

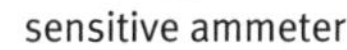

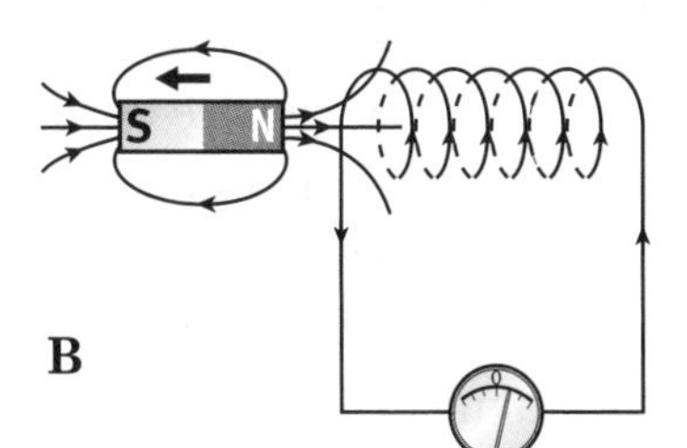

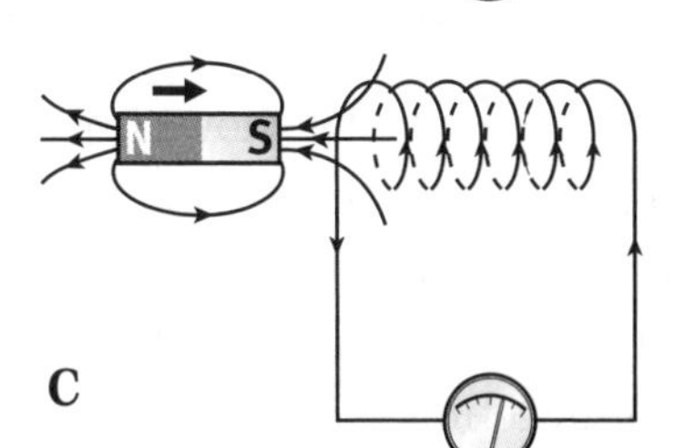

P5.4.1–2, P5.4.11–16 How is mains electricity generated?

Generators make electricity by electromagnetic induction. In a generator, a magnet or electromagnet turns near a coil of wire.

This induces a voltage across the ends of the coil.
The direction of this voltage changes each time the magnet rotates.

You can increase the size of the induced voltage by:

- turning the magnet faster
- making the magnetic field stronger
- adding more turns to the coil
- putting an iron core inside the coil.

Alternating current

The magnet in a generator turns all the time. Its magnetic field constantly changes direction. So the direction of the induced current changes all the time. This is an **alternating current (a.c.)**. The current from a battery is always in the same direction. It is a **direct current (d.c.)**.

Mains electricity is supplied as an alternating current (a.c.).

This is because:

- it is easier to generate than d.c.
- it can be distributed more efficiently, with less energy wasted as heat.

In the UK, domestic mains electricity is supplied at 230 volts.

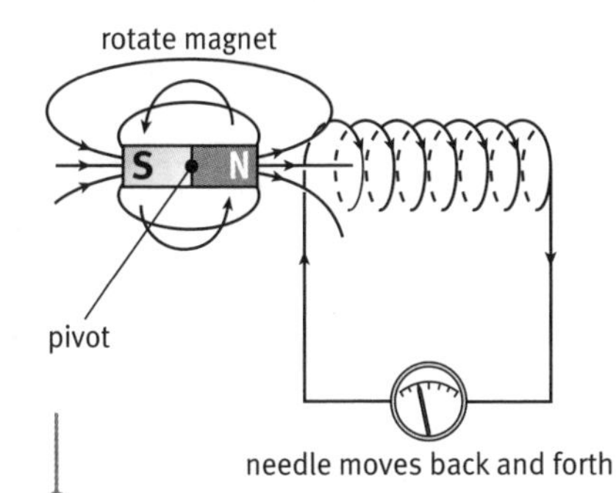

A generator.

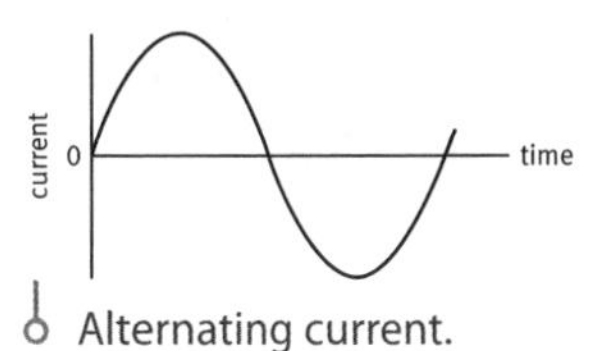

Alternating current.

P5.4.6–10 What are transformers?

If the current in a coil of wire changes, its magnetic field also changes. The changing magnetic field induces a voltage in a nearby coil.

A transformer consists of two coils of wire wound onto an iron core. It changes the size of an alternating voltage.

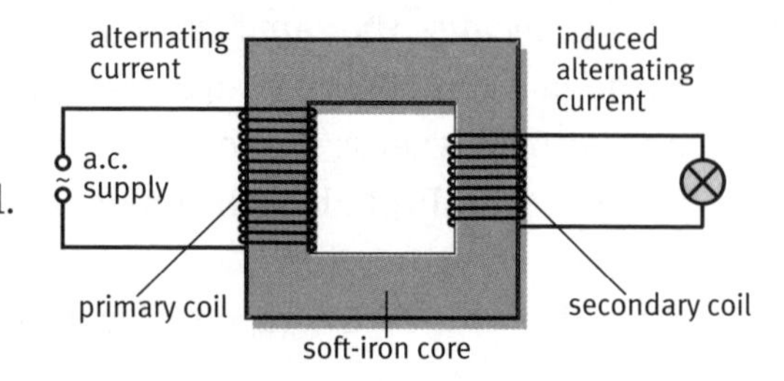

H The changing current in one coil of the transformer causes a changing magnetic field in the iron core. This induces a changing potential difference across the other transformer coil.

You can use this equation to work out the size of the voltage across the secondary coil:

$$\frac{\text{voltage across primary coil}}{\text{voltage across secondary coil}} = \frac{\text{number of turns on primary coil}}{\text{number of turns on secondary coil}}$$

P5.5.1–6 How are electric motors made?

If a current is flowing through a wire or coil, the wire or coil can exert a force on:

- a nearby permanent magnet
- another current-carrying coil or wire nearby.

For a current-carrying wire in a magnetic field:

- If the magnet's lines of force are at 90° to the wire, the wire experiences a force at 90° to both the current direction and the lines of force of the field.
- If the magnet's lines of force are parallel to the wire, the wire experiences no force.

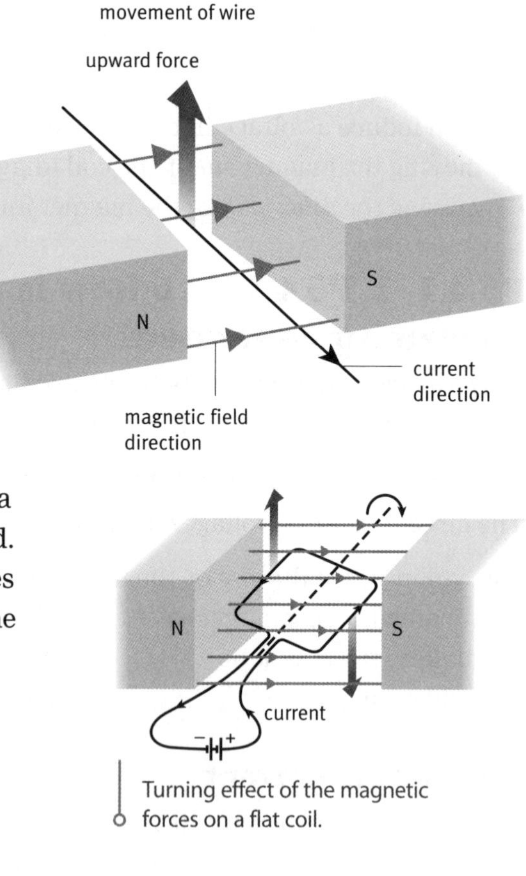

Turning effect of the magnetic forces on a flat coil.

The diagram opposite shows how the **motor effect** results in a turning force on a rectangular coil in a uniform magnetic field. If you add a **commutator** to the setup opposite, the coil rotates continuously. This happens because the commutator swaps the current direction every time the coil is vertical.

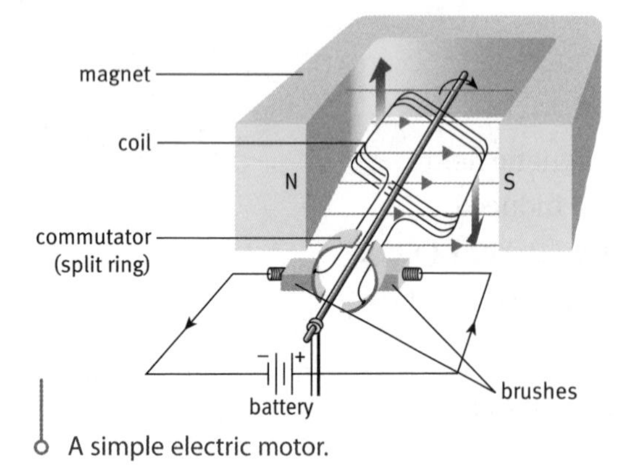

A simple electric motor.

Motors are used in computer hard disk drives, DVD players, and electric motor vehicles.

Use extra paper to answer these questions if you need to.

1 Draw the symbols for the components below.

a ammeter
b voltmeter
c cell
d power supply
e filament lamp
f switch
g light-dependent resistor (LDR)
h fixed resistor
i variable resistor
j thermistor

2 Use the sentence beginnings and endings to write eight full sentences.

Beginnings	Endings
All conductors	do not conduct electricity.
Metal conductors	include polythene, wood, and rubber.
Insulators	charges are not used up.
In a complete circuit	the battery makes free charges flow in a continuous loop.
	contain charges that are free to move.
	contain electrons that are free to move.
	do not contain charges that are free to move.

3 Highlight the correct word in each pair of **bold** words.
Resistors get **colder/hotter** when electric current flows through them. This is why lamp filaments glow.
The resistance of a light-dependent resistor changes with light intensity. Its resistance in the dark is **less/more** than its resistance in the light. The resistance of a thermistor changes with temperature. Usually, the higher the temperature, the **smaller/bigger** the resistance.

4 Write **X** next to the statements that are true for circuit X below. Write **Y** next to the statements that are true for circuit Y. Write **B** next to the statements that are true for both circuits.

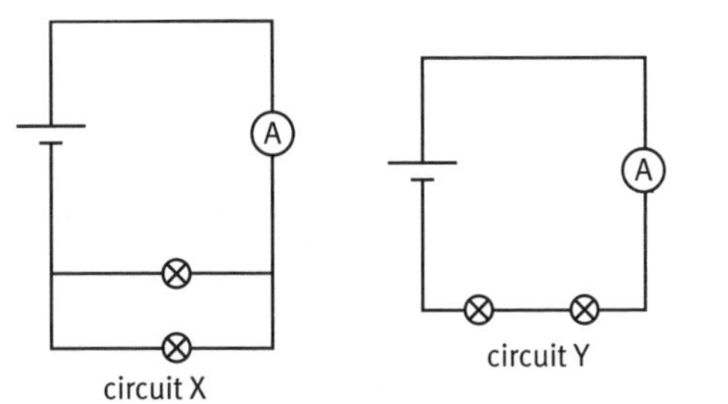

a There are several paths for the current.
b This circuit has a greater total resistance.
c The ammeter reading is smaller for this circuit.
d The components resist the flow of charge.
e The resistance of the connecting wires is so small that you can ignore it.
f It is easier for the battery to push charges around this circuit.

5 Fill in the gaps.

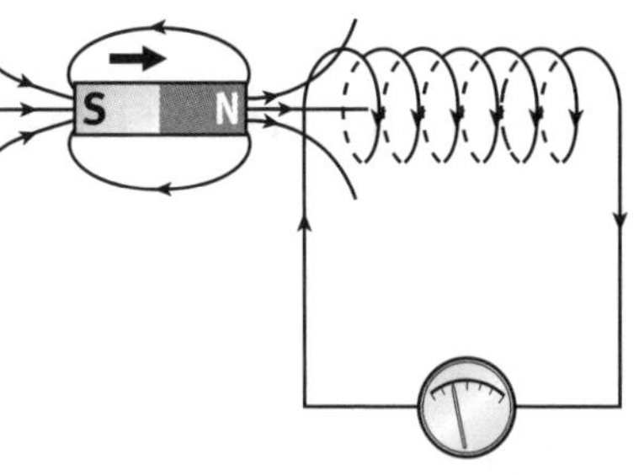

If you move the magnet into the coil of wire, a voltage is induced across the ends of the __________. If you connect the ends of the coil to make a closed circuit, a __________ flows round the circuit.
You can induce a voltage in the opposite direction by moving the magnet __________ of the coil or by moving the other __________ of the magnet into the coil.

6 A graph shows how the voltage produced by a generator changes with time when the magnet spins at a particular speed.

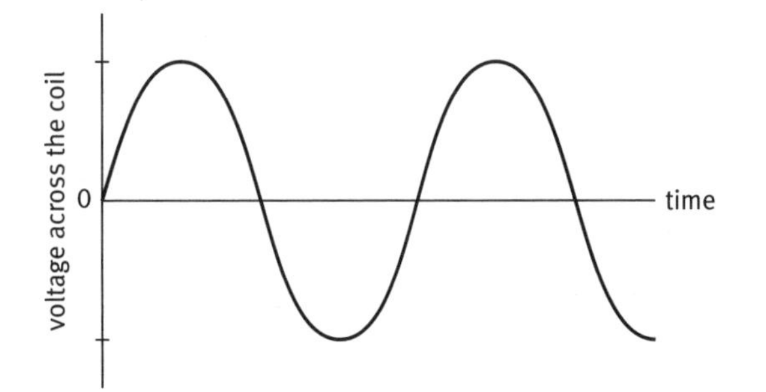

Sketch a graph to show what happens when the magnet is spun around more quickly.

7 Calculate the power of these appliances. Assume the voltage is 230 V.

a a vacuum cleaner with a current of 4 A flowing through it
b a DVD player with a current of 0.9 A flowing through it

H 8

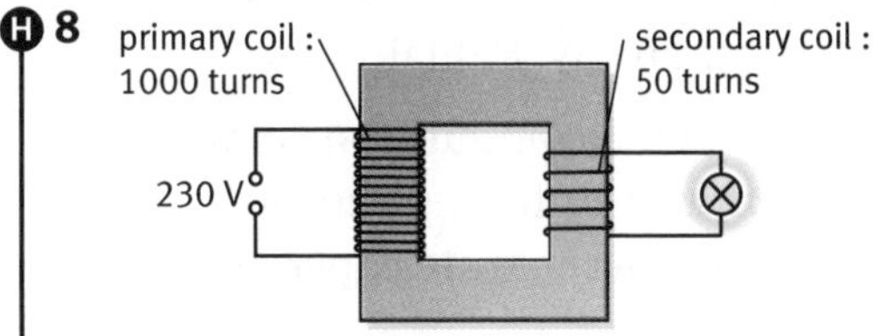

Calculate the voltage across the secondary coil.

9 Explain why:

a Resistors get hotter when an electric current flows through them.
b In a series circuit, the potential difference is largest across the component with the greatest resistance.
c Mains electricity is supplied as a.c.

1 Tamara has a portable heater. She plugs it into a car battery.

She puts the heating element into a mug of water to make a hot drink.

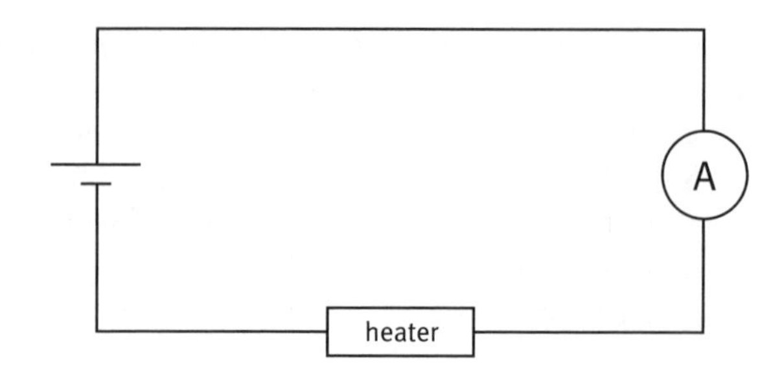

a Tamara wants to find out more about her heater.

She connects this circuit.

i Tamara uses a voltmeter to measure the voltage across the heater.

Draw on the diagram to show where to connect the voltmeter.

Use the correct symbol. [1]

ii The reading on the voltmeter is 12 V. The ammeter reads 10 A.

Calculate the resistance of the heater.

Resistance = ______________ ohms [2]

b The heater contains a heating element made from a coil of wire.

The wire gets hotter when an electric current passes through it.

Explain why the wire gets hotter.

____

____ [2]

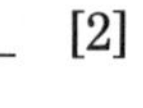

Total [5]

2 When a current flows in the circuit opposite, the coil rotates continuously. Explain why.

✎ The quality of written communication will be assessed in your answer to this question. Write your answer on separate paper or in your exercise book.

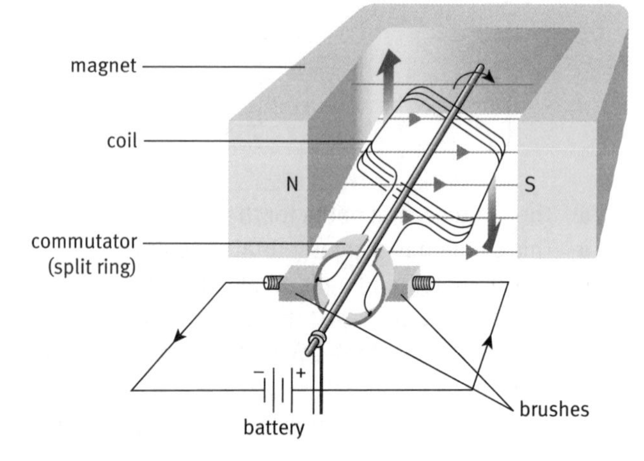

Total [6]

3 Scientists have invented a wind-up laptop computer. School students will use it in places where electricity supplies are not reliable.

A person turns a handle for 1 minute. This winds up a spring.

Then the spring unwinds slowly. This rotates a magnet within a coil of wire.

This generator produces an electric current.

a Suggest three changes the scientists could make to the generator so that it produced a bigger current.

Change 1: ____________________

Change 2: ____________________

Change 3: ____________________ [3]

b The computer can also be plugged into the mains electricity supply.

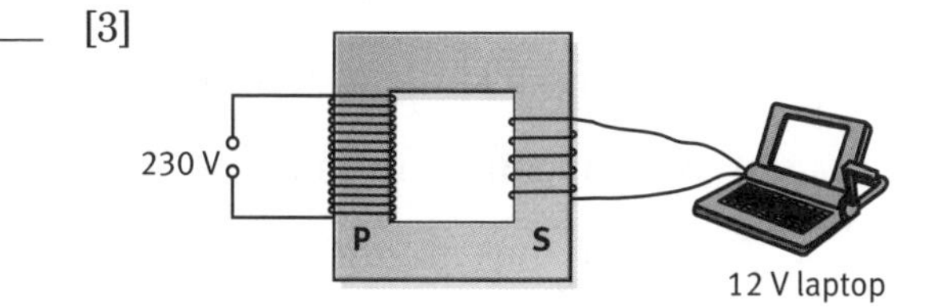

A transformer changes the size of the voltage.

i Explain how the transformer works.

____

____

____________________ [2]

ii Coil P has 2300 turns.

Calculate the number of turns needed in coil S to give a voltage of 12 V across the computer.

Answer = ____________ [2]

Total [7]

4 The diagram shows part of an electric circuit in Matt's house.

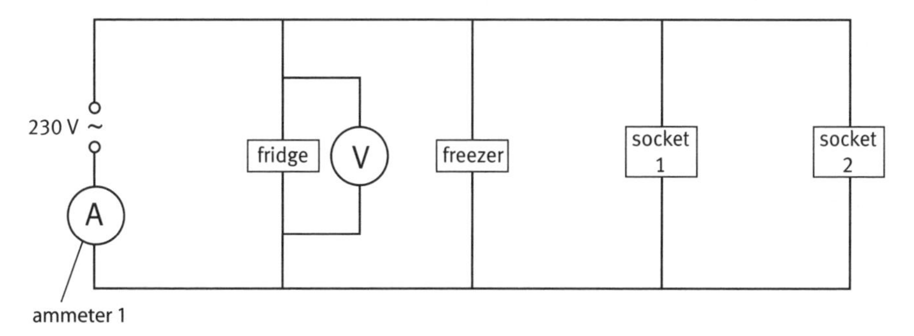

H

a i What is the reading on the voltmeter connected across the fridge?

_______________ [1]

ii What is the potential difference across the fridge?

_______________ [1]

b The resistance of the freezer is 70 Ω.

The voltage across the freezer is 230 V.

Calculate the current through the freezer.

Current = _______________ amps [2]

c Matt plugs a kettle into socket 1.

What happens to the size the current through the freezer?

Draw a (ring) around the correct answer.

increases **decreases** **stays the same** [1]

d Matt plugs a kettle into socket 1 and an electric heater into socket 2.

He switches off the freezer.

He measures the currents through the appliances that are switched on.

Appliance	Current (A)
fridge	0.4
kettle	5.0
heater	9.0

i What current flows through ammeter 1?

Current = _______________ amps [2]

ii Which appliance in the table has the greatest resistance?

_______________ [1]

H

Give a reason for your answer.

____

____ [1]

Total [9]

5 Vanessa has four resistors.

She sets up an experiment to measure the resistance of each one.

She records one value of voltage and one value of current for each resistor.

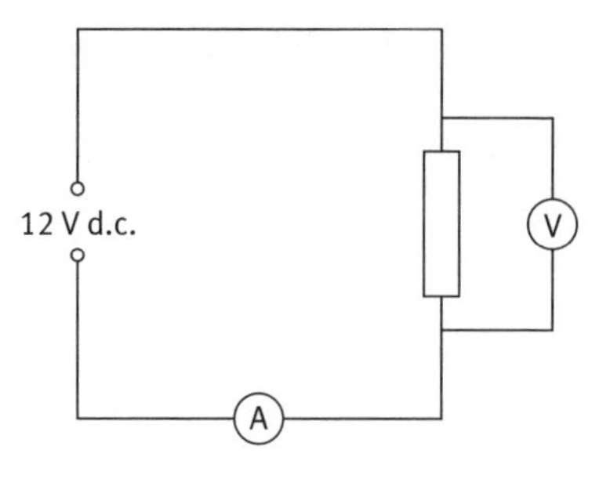

Resistor	Voltage (V)	Current (A)	Resistance (Ω)
J	12	4	
K	12	0.4	
L	12	0.2	
M	12	6	

a Use data from the table to calculate the resistance of each resistor.
Write your answers in the table above. [2]

b Vanessa gives Ursula one of her four resistors.

Ursula records the current that flows through the resistor at each of five different voltages.

She plots the values on the graph opposite.

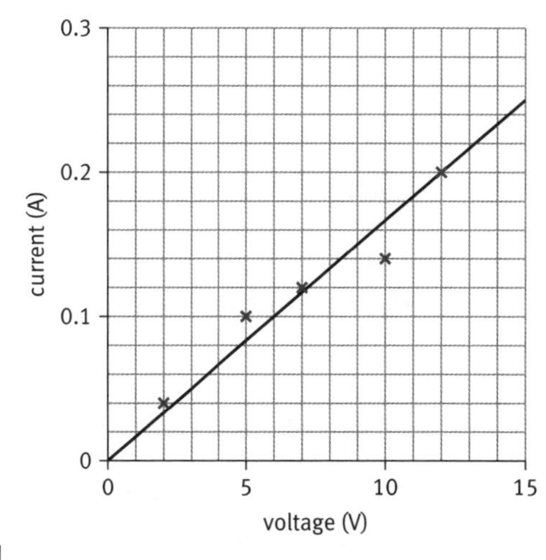

i Use the graph to calculate the resistance of the resistor.

Use the equation below and show your working.
Include the unit in your answer.

$$\text{resistance of resistor} = \frac{1}{\text{gradient of the graph}}$$

______________________________ [2]

ii From which resistor (J, K, L, or M) did Ursula obtain the data shown on the graph?

______________________________ [1]

c The graph opposite is for resistor M.

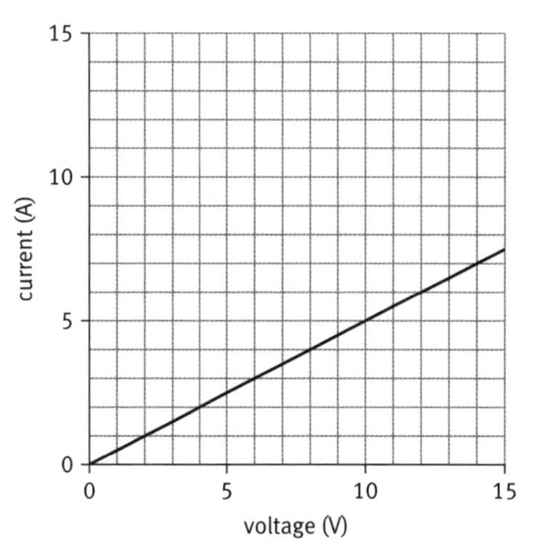

Explain how the graph increases or decreases your confidence that the resistance you calculated for resistor M in part is correct.

____

______________________________ [2]

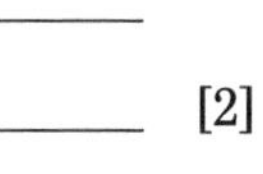

Total [7]

Going for the highest grades

H 6 Mary sets up the circuit opposite.

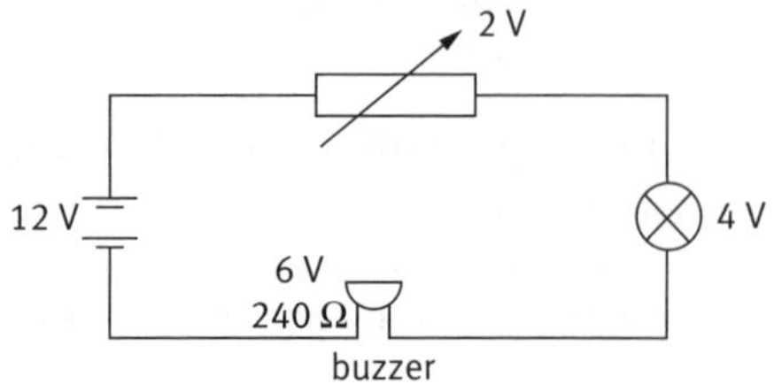

a What is the current through the lamp?

______________________ [1]

b Explain why the total potential difference across the three components of the circuit add up to the potential difference across the battery.

____

____

____

______________________ [2]

c Which component in the circuit has the greatest resistance?

Explain how you decided.

____

______________________ [1]

d Mary increases the resistance of the variable resistor.

What effect does this change have on the potential difference across the buzzer?

______________________ [1]

e Edward sets up a new circuit with the same components.

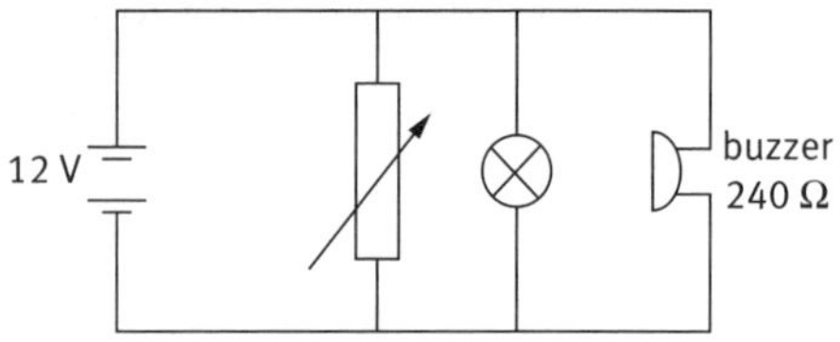

He moves the slider on the variable resistor back to its original position, before Mary moved it.

i Through which component does the largest current flow?

______________________ [1]

ii Explain why the current through the component you identified in part i is greater than the currents that flow through the other components in Edward's circuit.

____

______________________ [2]

Total [8]

2 V
12 V
4 V
6 V
240 Ω
buzzer

12 V
buzzer
240 Ω

P6 Radioactive materials workout

1 The diagram compares three types of ionising radiation.

For each bold pair of words or phrases, highlight the word or phrase that is correct.

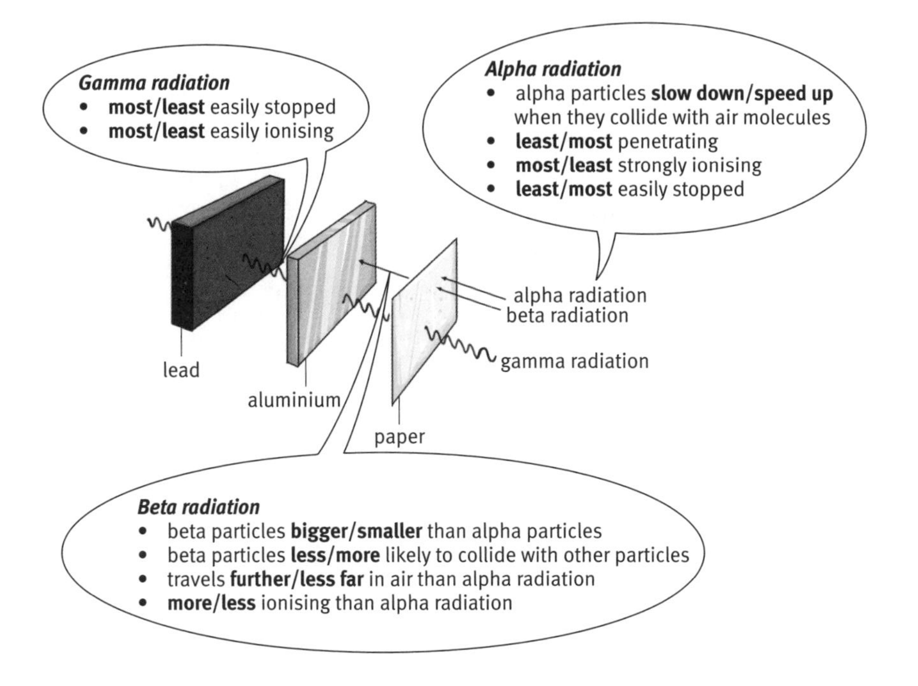

Gamma radiation
• most/least easily stopped
• most/least easily ionising
Alpha radiation
• alpha particles slow down/speed up when they collide with air molecules
• least/most penetrating
• most/least strongly ionising
• least/most easily stopped
alpha radiation
beta radiation
gamma radiation
lead
aluminium
paper
Beta radiation
• beta particles bigger/smaller than alpha particles
• beta particles less/more likely to collide with other particles
• travels further/less far in air than alpha radiation
• more/less ionising than alpha radiation

2 The table shows how different sources of radiation contribute to the average radiation dose in the UK.

Write the name of each radiation source by the correct section of the pie chart.

Radiation source	Average radiation dose in the UK (mSv)
radon gas from the ground	1.25
food and drink	0.24
cosmic rays	0.30
medical	0.38
gamma rays from the ground and buildings	0.33
fallout	0.0005
occupational	0.0005

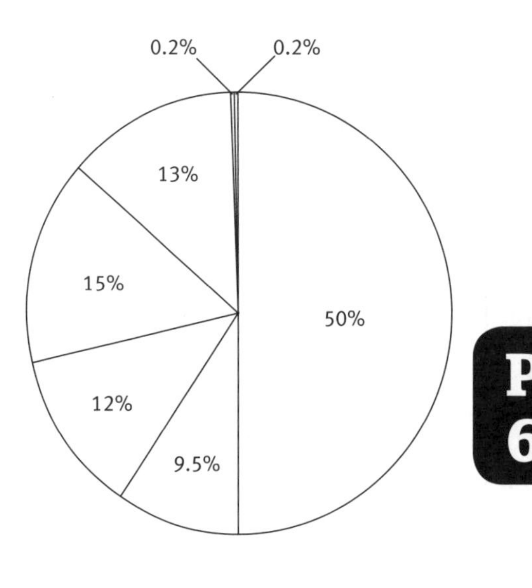

P6

3 Fill in the gaps to complete the labels on this diagram of an atom.

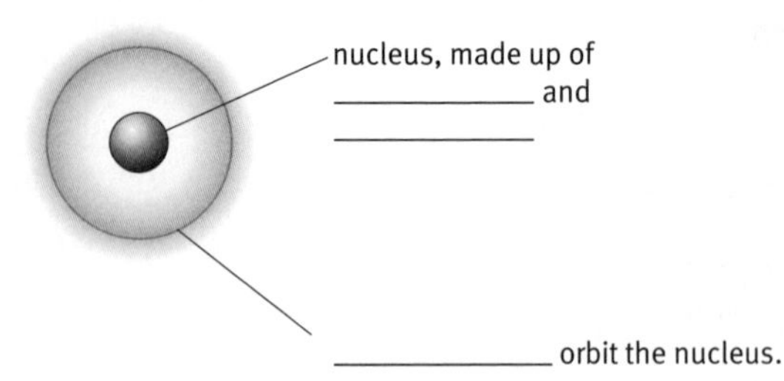

4 Annotate the diagram to describe and explain what happened when Geiger and Marsden fired alpha particles at a piece of gold foil in a vacuum.

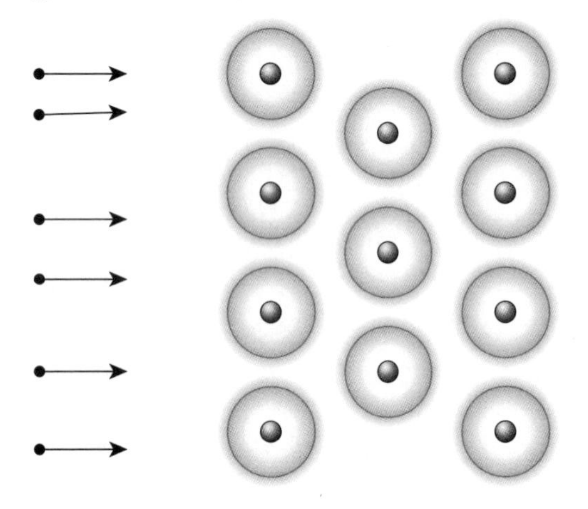

H 5 The diagrams show the number of protons and neutrons in six atoms.

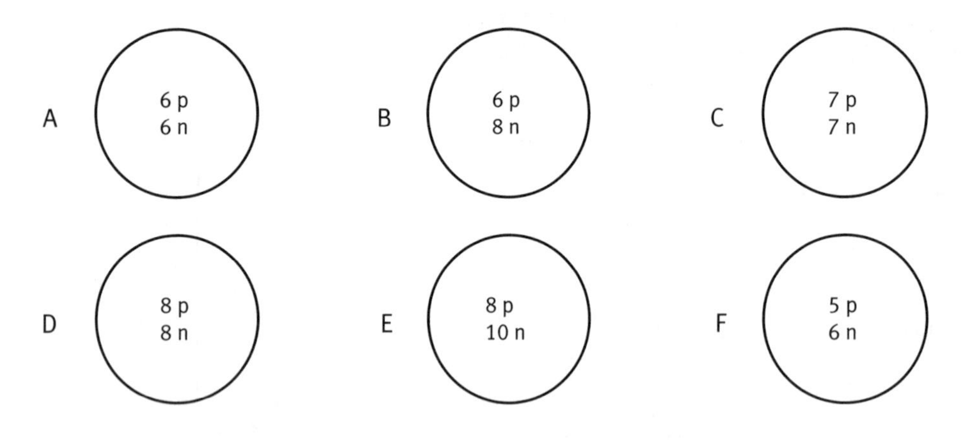

a Give the letters of two pairs of atoms of the same element.

____

b Give the letter of the atom that has the fewest total number of particles in its nucleus. _________

c Give the letter of the atom that has the greatest total number of particles in its nucleus. _________

d Give the letters of two atoms that have the same total number of particles in their nuclei. _______________

P6.1.1, P6.1.9 What are radioactive materials?

Radioactive materials give out (emit) ionising radiation all the time. You cannot change the behaviour of a radioactive material – it emits radiation whatever its state (solid, liquid, gas) and whether or not it has taken part in a chemical reaction.

P6.1.10–11 What types of ionising radiation are there?

Type of radiation	Penetration properties	H What is it?
alpha (α)	absorbed by paper, clothing, skin, and a few cm of air	a positively charged particle made up of two protons and two neutrons
beta (β)	penetrate paper; absorbed by a thin sheet of metal	an electron
gamma (γ)	absorbed only by thick sheets of dense materials (e.g. lead) or several metres of concrete	a high-energy electromagnetic wave

P6.1.3–6 What's in an atom?

Three scientists (Geiger, Marsden, and Rutherford) fired alpha particles at thin gold foil in a vacuum. Most alpha particles passed through the foil. A few of the positively charged alpha particles were reflected backwards.

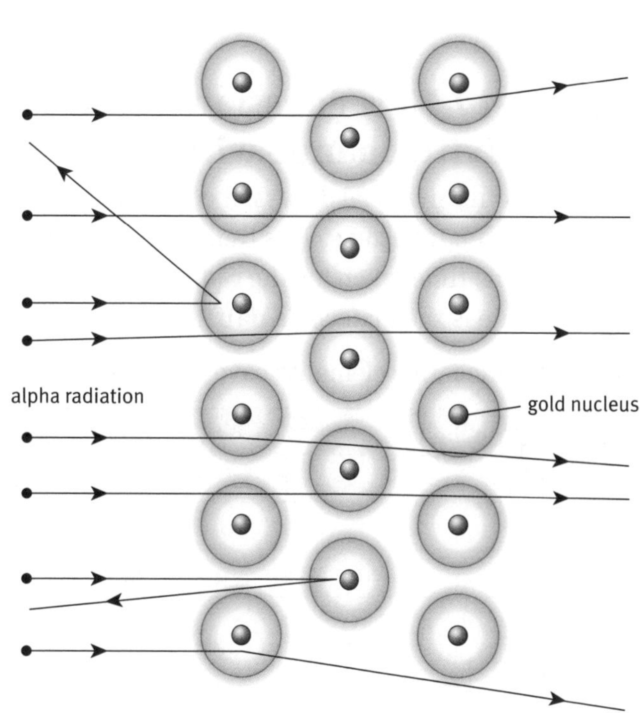

The scientists concluded that a gold atom has a small, massive, positive region at its centre – its **nucleus**.

We now know that every atom has a tiny core, or **nucleus**. The nucleus is surrounded by **electrons**.

The nucleus is made up of **protons** and **neutrons**. Protons are positively charged. Neutrons have no charge.

H Protons and neutrons are held together by a strong force. This balances the repulsive electrostatic force between the protons.

If two hydrogen nuclei are brought closely together, they may join to make a helium nucleus. The process releases energy. It is called **nuclear fusion**.

H P6.1.7–8 and P6.1.12–13 What makes a substance radioactive?

Every atom of a certain element has the same number of protons. Different atoms of this element may have different numbers of neutrons. Atoms of the same element that have different numbers of neutrons are called **isotopes**. For example:

Name of isotope	Number of protons	Number of neutrons	Is this isotope radioactive?
carbon-12	6	6	no
carbon-14	6	8	yes

The nucleus of carbon-12 is stable. It is not radioactive.

The nucleus of carbon-14 is unstable. It decays to make a stable nucleus of another element, nitrogen. As it decays, it emits beta radiation. You can summarise this reaction in a **nuclear equation**:

$$^{14}_{6}\mathrm{C} \longrightarrow ^{14}_{7}\mathrm{N} + ^{0}_{-1}\beta$$

Other elements have atoms with unstable nuclei, for example, radium-226. When a radium-226 atom decays, it emits alpha radiation to make a radon-222 atom. This is the decay product.

$$^{226}_{88}\mathrm{Ra} \longrightarrow ^{222}_{86}\mathrm{Rn} + ^{4}_{2}\alpha$$

Exam tip

Practise writing nuclear equations.

When an atom decays, energy is released. Einstein's equation calculates how much energy is released for a given loss of mass.

energy = change in mass × (speed of light)^2

or $E = mc^2$

P6.1.14–17 How does a material's radioactivity change with time?

As a radioactive material decays, it contains fewer atoms with unstable nuclei. It becomes less radioactive and emits less radiation. The time taken for the radioactivity to fall to half its original value is the material's **half-life**.

Different radioactive elements have different half-lives:

Radioactive element	Half-life
plutonium-242	380 thousand years
carbon-14	5.6 thousand years
strontium-90	28 years
iodine-131	8 days
lawrencium-257	8 seconds

The shorter the half-life, the greater the activity for the same amount of material.

The graph shows the decay curve for iodine-131. After 8 days, 50% (half) of the original sample remains, and the activity is half its original value.

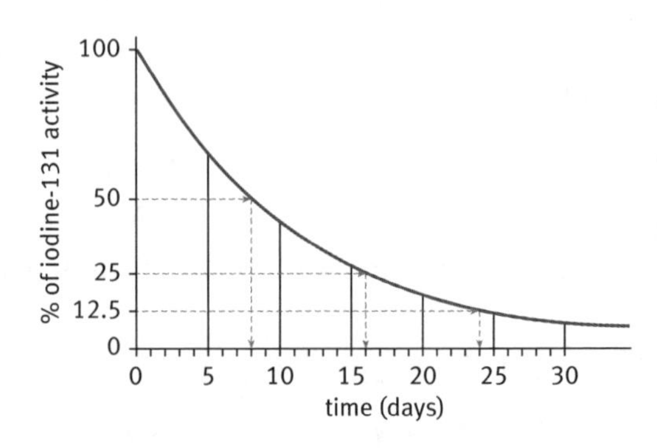

P6.1.2, P6.2.1–2, P6.2.6–7 What are the risks from radioactive sources?

We are exposed to **background radiation** all the time. There are many radioactive substances in the environment. These contribute to background radiation.

If ionising radiation reaches you from a source outside your body, you are being **irradiated**. If a radioactive material gets onto your skin or clothes, or inside your body, you are **contaminated**. You will be exposed to the radiation as long as the material stays there.

When ionising radiation hits atoms and molecules, it may break them into charged particles, called **ions**.

The ions formed may then take part in other reactions.

The ions damage living cells:

- Larger amounts of radiation may kill cells.
- Smaller amounts may damage a cell's DNA, causing cancer.

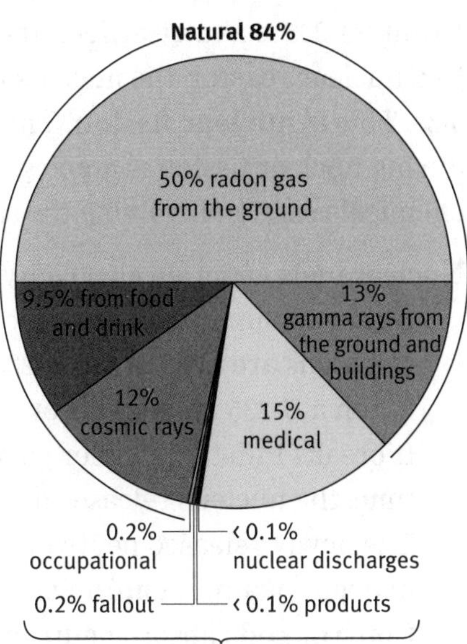

The UK average annual dose is 2.5mSv.

This pie chart shows the contribution of different sources to the average background radiation dose in the UK. Source HSE.

P6.2.3 How is ionising radiation useful?

Ionising radiation has many uses:

- To treat cancer by **radiotherapy**.

  The ionising radiation damages cancerous cells and they stop growing.

- To sterilise surgical instruments, and herbs and spices.

  The ionising radiation kills bacteria on the instruments or food.

- To help diagnose disease by acting as a tracer in the body.

  For example, radioactive krypton-81m gas shows doctors how gases move in diseased lungs.

P6.2.4, P6.2. 9 What are the risks of handling radioactive materials?

The more ionising radiation a person is exposed to, the greater the risk to health. **Radiation dose** measures the possible harm to your body. It takes account of the amount and type of radiation. Its units are sieverts (Sv).

Hospital radiographers and nuclear power station workers are exposed to radioactive sources. Their exposure is monitored.

P6.2.10–13 How do nuclear power stations generate electricity?

Radioactive materials release energy from changes in the nucleus. They can be used as **nuclear fuels**.

Uranium-235 is a nuclear fuel. Its nucleus is large and unstable. In a nuclear reactor, the nucleus breaks into two parts of similar size. This is **nuclear fission**. The amount of energy released during nuclear fission is much greater than that released in a chemical reaction involving the same mass of material.

H Nuclear fuels generate electricity like this:

- **Fuel rods** in the nuclear reactor contain uranium-235 (U-235).
- Neutrons are fired at the U-235.
- When a U-235 nucleus absorbs a neutron, it becomes unstable. It breaks into two smaller parts of similar size. At the same time, the nucleus releases more neutrons. Energy is released.
- The newly released neutrons hit more U-235 nuclei. Fission happens again. A **chain reaction** has started.
- **Control rods** absorb neutrons. They are moved into or out of the reactor to control the reaction rate.
- A fluid **coolant** is pumped through the reactor. The fuel rods heat up the coolant.
- The coolant heats up water, which becomes steam.
- The steam turns turbines, which turn a generator.

P6.2.8, P6.2.14–15 What happens to nuclear waste?

Nuclear power stations produce dangerous radioactive waste.

Scientists use half-lives to work out when nuclear waste will become safe. Elements that have long half-lives remain hazardous for many thousands of years; those with short half-lives quickly become less dangerous.

Type of waste	Example	How it is disposed of
low level	used protective clothing	packed in drums and dumped in a lined landfill site
intermediate level	materials that have been inside reactors – may remain highly radioactive for many years	mixed with concrete and stored in stainless steel containers
high level	concentrated radioactive material from spent fuel rods	decays fast and releases energy rapidly, so needs cooling; in the UK, stored in a pool of water at Sellafield

Use extra paper to answer these questions if you need to.

1 In the list below, tick the statements that describe ways in which ionising radiation is used in the UK.
- a to treat cancer ☐
- b as a tracer in the body ☐
- c to sterilise spices ☐
- d to generate electricity ☐

2 Write **T** next to the statements that are true. Write corrected versions of the statements that are false.
- a Radioactive materials emit radiation all the time.
- b Atoms of carbon-14 are radioactive. If a carbon-14 atom joins to oxygen atoms to make carbon dioxide, the carbon dioxide will not be radioactive.
- c Solid caesium chloride that is made with caesium-137 is radioactive. It remains radioactive when it dissolves in water.
- d Radiation dose is measured in half-lives.
- e Radiation dose is based on the amount and type of radiation a person is exposed to.
- f Hydrogen nuclei can join together to make helium nuclei. This process is called nuclear fission.
- g The energy released in a nuclear reaction is much less than the energy released in a chemical reaction involving a similar mass of material.

3 Draw lines to match each word or phrase to its definition.

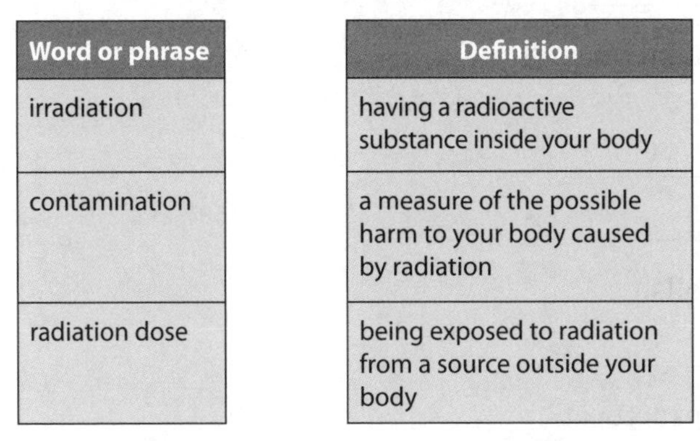

Word or phrase	Definition
irradiation	having a radioactive substance inside your body
contamination	a measure of the possible harm to your body caused by radiation
radiation dose	being exposed to radiation from a source outside your body

4 List two groups of people whose exposure to radiation is carefully monitored.

5 Doctors use radioactive krypton-81m gas to help diagnose lung diseases.
- a Krypton-81m has a half-life of 13 seconds. Explain why it would be a problem if the half-life was much longer than this.
- b A sample of krypton-81m has an activity of 4000 Bq. After how long will its activity be reduced to 1000 Bq?
- c Krypton-81m emits gamma rays. Suggest two safety precautions hospital staff must take when using this gas.

H 6 Complete the sentences below.

In the nucleus of an atom, protons and _______ are held together by a _______ force. This balances the repulsive _______ force between the protons.

A radioactive atom has an _______ nucleus. The nucleus _______ to become more stable, emitting energetic _______ in the process.

7 The table shows the half-lives of some radioactive uranium isotopes.

Element	Half-life
uranium-222	1.4 milliseconds
uranium-228	9 minutes
uranium-231	4 days
uranium-232	69 years

- a A scientist has some uranium-228. Its relative activity is 100. After what time period will its relative activity be 50?
- b A scientist has some uranium-231. After what time period will its activity be one quarter that of its original activity?
- c A scientist has some uranium-232. Its relative activity is 100. After what time period will its relative activity be 12.5?

8 Give the numbers of protons and neutrons in each of the isotopes of uranium below.

a ${}^{239}_{92}U$ b ${}^{225}_{92}U$ c ${}^{217}_{92}U$

9 A nuclear power station releases 66×10^9 J of energy to provide a family of four with their electricity needs for one year. Calculate the mass of fuel that must be lost to provide this energy.

Use the equation $m = \frac{E}{c^2}$.

The value of c is 3×10^8 m/s.

10 Describe the purpose of each of the following in a nuclear power station.

a fuel rod b control rod c coolant

11 Complete the nuclear equations by giving the proton number and atomic mass/mass number for each of the atoms, alpha particles, and beta particles.
- a ${}^{239}_{92}U \longrightarrow Np + {}_{-1}\beta^{0}$
- b ${}^{14}C \longrightarrow {}_{7}N + \beta$
- c ${}^{235}U \longrightarrow {}_{90}Th + {}_{2}\alpha^{4}$
- d ${}^{209}_{83}Bi \longrightarrow {}^{205}_{81}Tl +$ _____
- e ${}_{86}Rn \longrightarrow {}^{215}Po + \alpha$

1 Read the article about treating cancer with radioactive materials.

> Arthur has a cancer tumour deep inside his body. His doctors will use radiotherapy to treat it. Arthur's doctors and radiotherapists plan the treatment carefully. They tattoo his skin to show exactly where to direct the radiation, and calculate the dose of radiation Arthur must receive.
>
> Arthur gets his treatment in a lead-lined room. When everything is ready, the radiotherapist leaves the room. Once outside, she switches on the treatment machine. Gamma rays enter Arthur's body for a few minutes. During the treatment, the radiotherapist watches Arthur on closed-circuit television. They can talk to each other over an intercom.
>
> Arthur goes to hospital for treatment every weekday for five weeks. On each visit, the gamma radiation enters his body at a different angle.

a **i** What type of material emits the radiation that enters Arthur's body?

____

ii Why does the radiotherapist use gamma radiation, and not alpha or beta radiation?

____

______________________________ [2]

b **i** What does gamma radiation do to cancer cells?

____

ii Why is it important to direct the radiation exactly at the cancer tumour?

____

______________________________ [2]

c **i** Why are the walls of the treatment room lined with lead?

____

ii Suggest why the radiotherapist leaves the room while Arthur is receiving his treatment.

____

______________________________ [2]

Total [6]

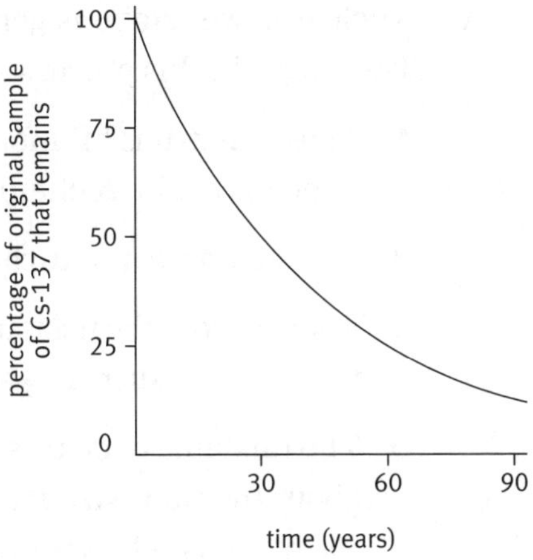

2 Caesium-137 (Cs-137) emits beta particles. It is used to treat some cancers.

The graph shows how the activity of this radioactive source changes over time.

percentage of original sample of Cs-137 that remains

100
75
50
25
0

30 60 90

time (years)

a Read the statements below.

Put ticks in the boxes next to each true statement.

The activity of the Cs-137 source decreases over time. ☐

All radioactive elements have a half-life of between 10 and 50 years. ☐

The half-life of Cs-137 is 30 years. ☐

The longer the half-life of a radioactive source, the more quickly it becomes safe. ☐

Beta radiation is absorbed only by thick sheets of lead or concrete. ☐ [2]

H

b A sample of caesium chloride contains 10 g of caesium-137.

Calculate the mass of caesium-137 that will remain after 120 years.

____

____ [2]

c Caesium-137 decays to barium-137. Barium-137 is not radioactive.

Complete the following sentences.

Use the words in the box.

negative	**unstable**	**stable**	**neutral**

The nucleus of a caesium-137 atom is ________________.

It decays and emits beta radiation. This makes barium-137,

which has a nucleus that is ________________. [2]

Total [6]

P6

3 a Nuclear power stations generate electricity.
The stages in this process are shown below.

A These neutrons hit more uranium-235 nuclei. Fission happens again. A chain reaction has started.

B The steam turns a turbine.

C Energy from the fission reaction is transferred as heat to a coolant, such as water or carbon dioxide.

D The unstable nucleus splits into two smaller parts of about the same size. This is fission. At the same time, the nucleus releases more neutrons.

E Neutrons are fired at fuel rods.

F When a neutron hits the nucleus of a uranium-235 atom, the nucleus becomes unstable.

G The hot coolant heats up water in a boiler to make steam.

The stages are in the wrong order.

Write a letter in each box to show the correct order.

E	F					B

[1]

b Complete the following sentences.
Choose from the words in the box.

barium	**protons**	**electrons**	**boron**
bismuth	**neutrons**	**rate**	

Control rods control the ____________ of fission reactions when they are lowered into or raised out of the nuclear reactor. They contain ____________ to absorb ____________. [3]

c Nuclear power stations produce radioactive waste.
Draw straight lines to match each **type of waste** to its **disposal method**.

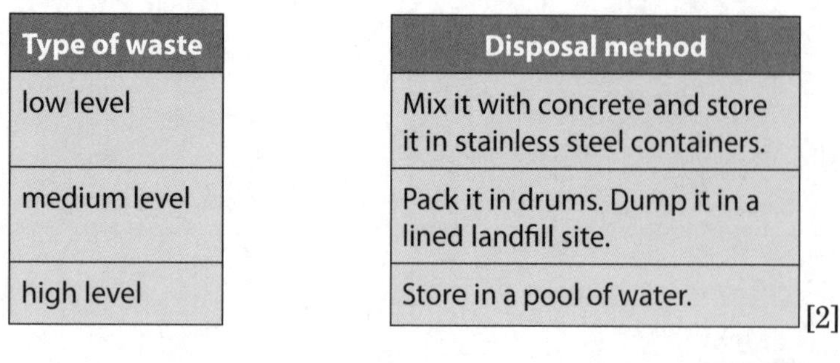

Type of waste
low level
medium level
high level

Disposal method
Mix it with concrete and store it in stainless steel containers.
Pack it in drums. Dump it in a lined landfill site.
Store in a pool of water.

[2]

Total [6]

4 Scientists asked parents in five US states to send in baby teeth from their children. The scientists measured the amounts of strontium-90 in about 2000 of these teeth. In 2003, scientists published a scientific paper about their findings.

The bar charts show some of their results.

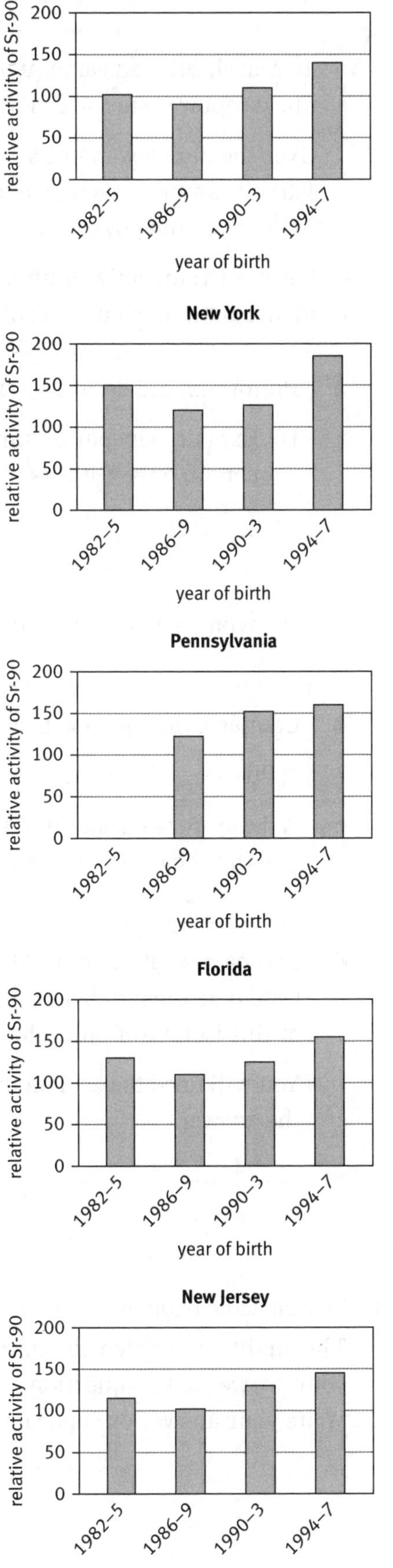

a Look at the bar charts.

i Describe the trend shown by the bar charts.

____

______________________________ [2]

ii Suggest why the graph for Pennsylvania includes no data for people born between 1982 and 1985.

______________________________ [1]

b The scientists suggested that the levels of strontium-90 were caused by an increase in the amount of electricity generated in nuclear power stations from 1986 onwards.

Do the data in the bar charts support this conclusion? Give a reason for your decision.

____

______________________________ [1]

c The scientists looked at their data again. They found that teeth from children living within 64 km of a nuclear power station had up to 54% more strontium-90 in them than teeth from children living further from nuclear power stations.

Does this finding make the conclusion in part more or less likely to be correct? Give a reason for your decision.

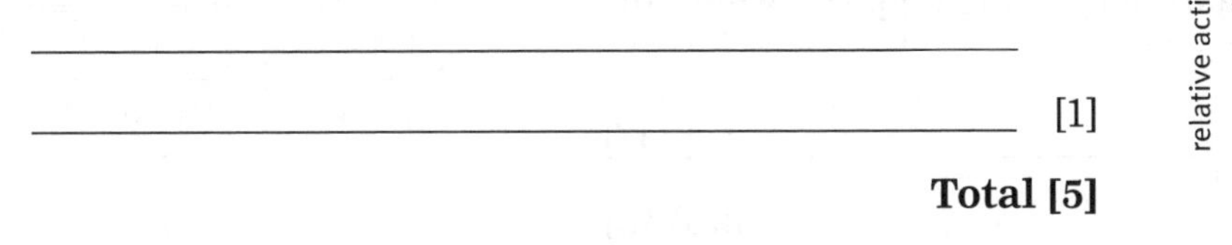

____

______________________________ [1]

Total [5]

P 6

Going for the highest grades

H 5

In March 2011, an earthquake damaged a nuclear power station in Fukushima, Japan.

Over the next few months, scientists analysed 100 soil samples taken from distances of up to 80 km from the power station.

Levels of radioactive plutonium-238 were higher than normal in six of the soil samples, including in a sample taken from a village 45 km from the power station.

Plutonium-238 from the soil can enter the body if it is breathed in, or eaten with food.

It is then deposited in the lungs and bones.

a Plutonium-238 decays by emitting alpha particles.

i Explain why plutonium-238 may be a health risk to a person only if it gets inside their body.

____

________________________________ [2]

ii What is an alpha particle made up of?

________________________________ [1]

b Complete the equation below for the decay of plutonium-238.

$^{238}_{94}\text{Pu} \longrightarrow \quad \text{U} + \quad \alpha$ [2]

c A decay curve from Pu-238 is shown below.
Use the decay curve to estimate the half-life of Pu-238.

________________________________ [2]

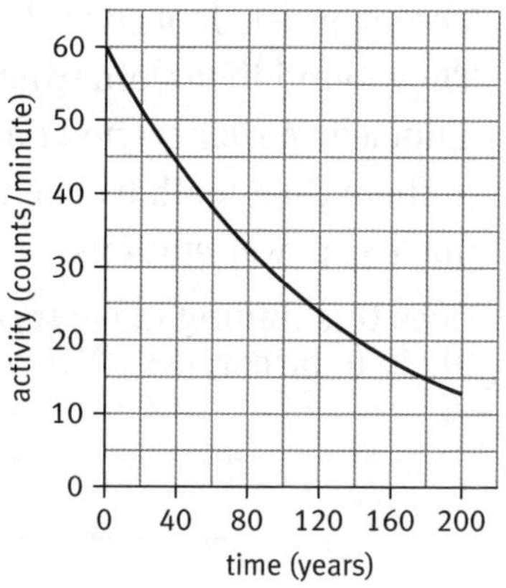

activity (counts/minute)
60
50
40
30
20
10
0
0 40 80 120 160 200
time (years)

d Radioactive plutonium-239 was found in some soil samples. Calculate the number of protons and neutrons in an atom of this isotope of plutonium.

You will need to use data from part **a** to help you work out the answer.

________________________________ [1]

Total [8]

H **6** Describe and explain how ionising radiation is used in hospitals.
The quality of written communication will be assessed in your answer to this question.
Write your answer on separate paper or in your exercise book.

Total [6]

P7.1 Naked-eye astronomy

workout

1 Match the descriptions of the Moon with its appearance and position.

Look at the diagram (positions 1–8) to help you complete the first column.

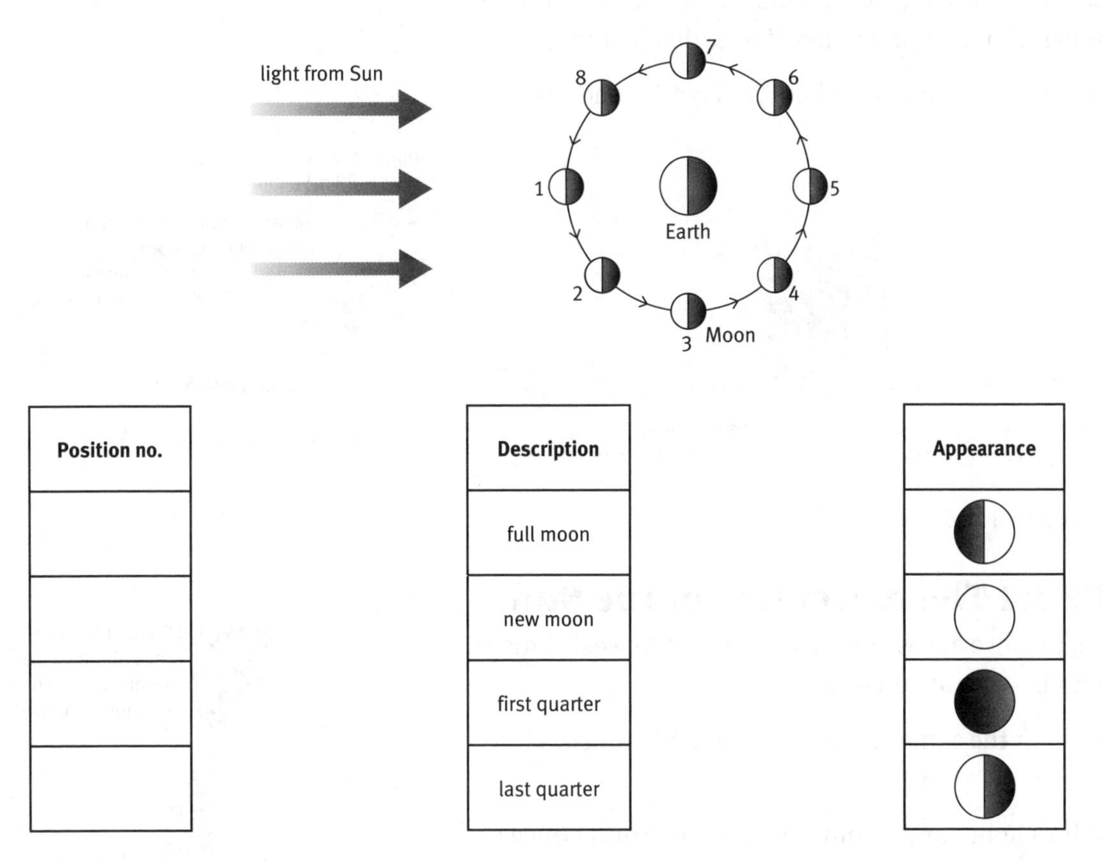

Position no.	Description	Appearance
	full moon	
	new moon	
	first quarter	
	last quarter	

H **2** The diagram below shows the positions of the Earth and the planet Mars at intervals of one month.

a Draw straight lines to show the direction in which Mars is seen against the background of the fixed stars.

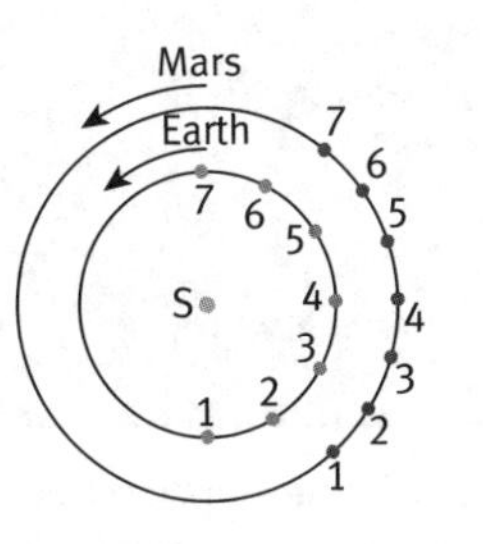

b Between which numbered months does Mars appear to move backwards? ____________

What is the Solar System?

The Solar System is the collection of planets, comets, and all other objects that **orbit** the Sun. The Sun is a star.

The Earth takes 365 ¼ days to complete one orbit. We call this one **year**. It also rotates about an imaginary line called its **axis**.

The Moon orbits the Earth. It takes about 28 days for one orbit.

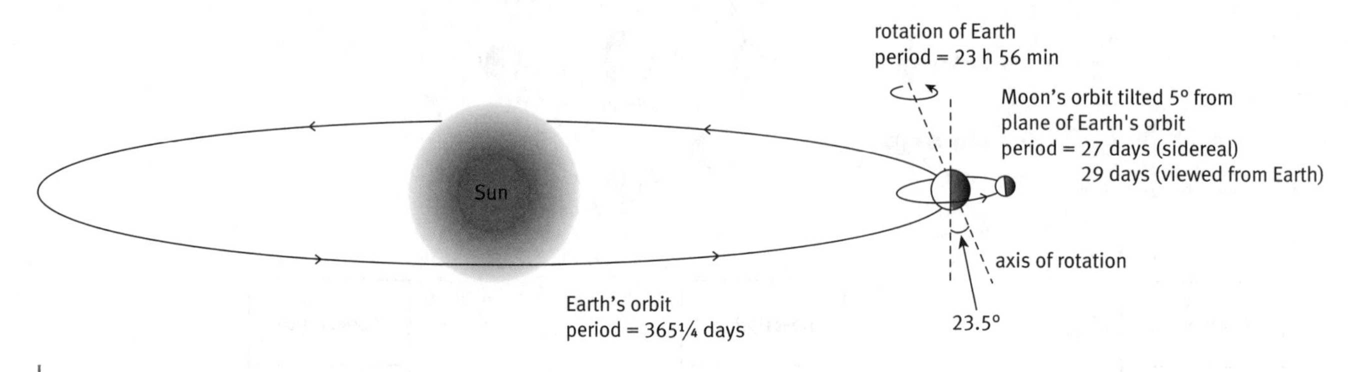

The orbits of the Earth and Moon (not to scale).

P7.1.1–2, P7.1.4 The movement of the Sun

The Sun appears to move across the sky from east to west. This is because the Earth is spinning on its axis.

The Sun reappears in the same place once every **24 hours**. This is a **solar day**.

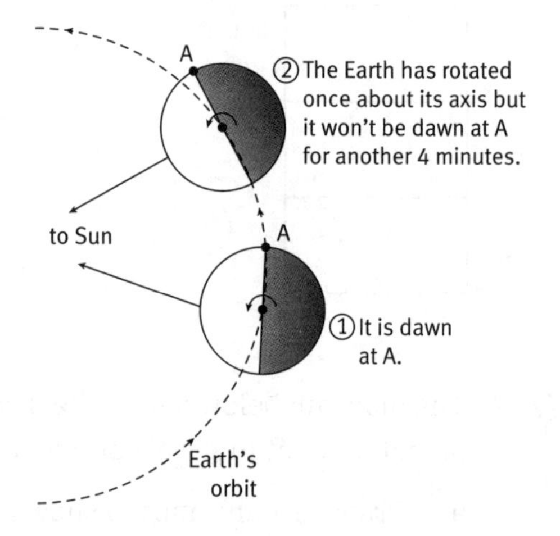

H It takes the Earth 23 hours and 56 minutes to spin around once on its axis, a rotation of 360°. This is a **sidereal day**.

During a sidereal day, the Earth also moves further along its orbit around the Sun. For the same part of the Earth to face the Sun again, it needs to turn for an extra 4 minutes. This explains why a solar day is 4 minutes longer than a sidereal day.

P7.1.1, P7.1.4, P7.1.7 The movement of the stars

Long-exposure photographs show the stars to be moving in circles around the Pole Star. Of course the stars are not actually moving like this – we are observing them from a spinning Earth. The stars will appear to be back in the same places after 23 hours and 56 minutes, when the Earth has rotated once about its axis.

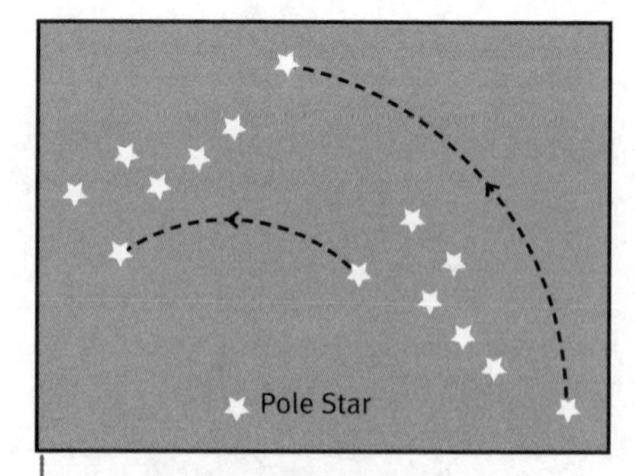

From the UK, the stars look as if they are moving in circles around the Pole Star.

A group of stars that form a pattern is a **constellation**. We see different constellations in summer and winter because of the Earth's movement around the Sun.

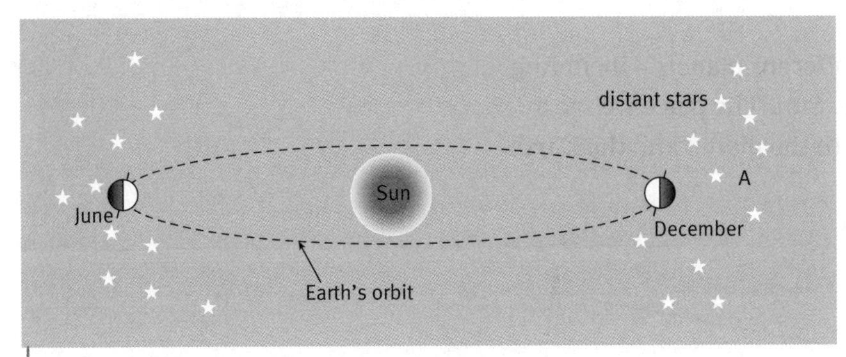

We can see the distant stars at A in December but not in June.

P7.1.1, P7.1.4 The movement of the Moon

The Moon appears to move across the sky from east to west, like the Sun. But the Moon takes longer – it reappears in the same part of the sky every 24 hours and 49 minutes.

The longer time for the Moon is explained like this:

- As well as the Earth's rotation giving us a different view of the Moon, the Moon itself is orbiting the Earth. One orbit takes about 28 days.
- The Moon orbits the Earth from west to east. So during the night the position of the Moon over 28 days appears to slip slowly back through the pattern of the stars.

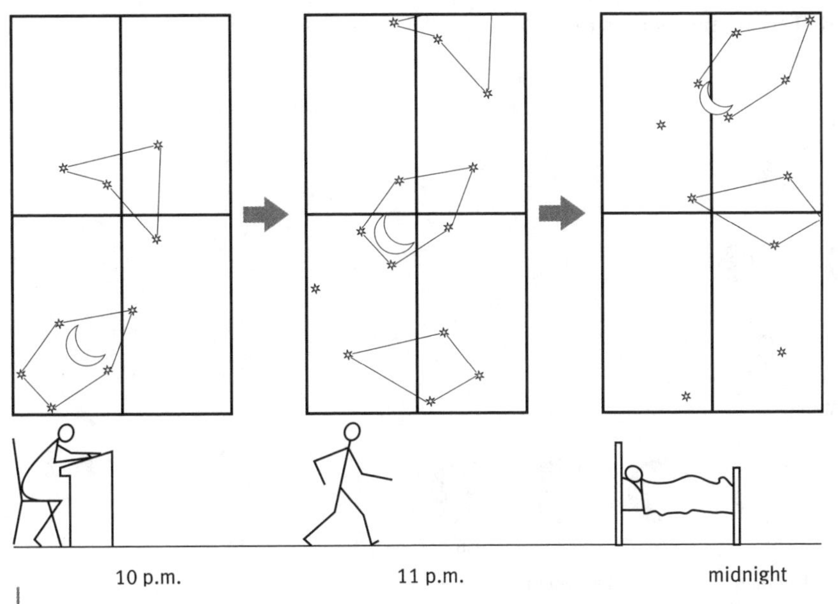

View through a window showing the Moon moving from east to west through the night, but slipping back relative to the pattern of stars.

P7.1.3–4, P7.1.8 Moving planets

You can see the planets Mercury, Venus, Mars, Jupiter, and Saturn with your naked eye. The planets orbit the Sun. This makes their positions appear to change night by night against the background of the fixed stars.

For most of the time the planets seem to move in a steady pattern across the sky, from east to west like the Sun and Moon. But sometimes the planets seem to go backwards. This is called **retrograde motion**.

Retrograde motion happens because different planets – including Earth – take different times to orbit the Sun. The place we see a planet in the sky depends on where both the planet and the Earth are in their orbits.

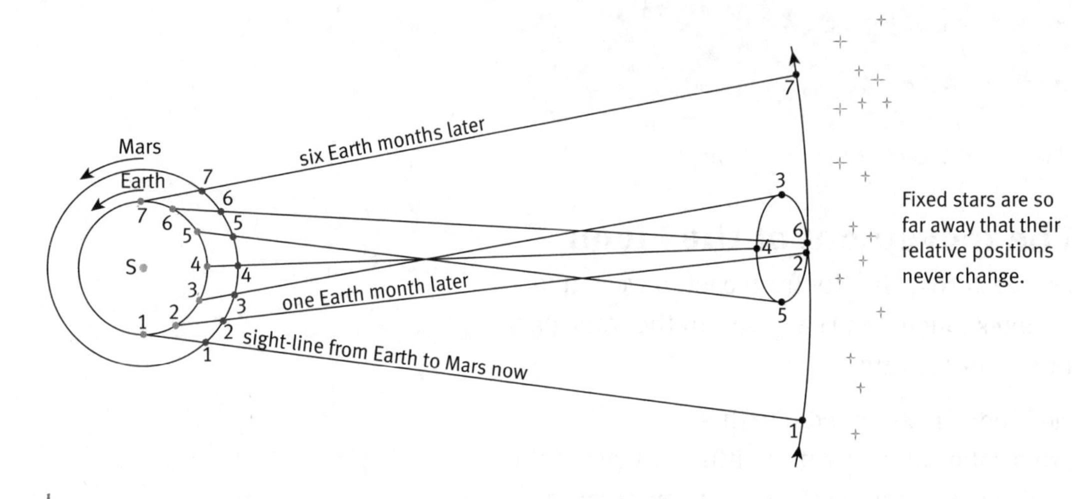

From months 1 to 3, Mars appears to move forwards. Then, for two months, it goes into reverse before moving forward again.

P7.1.5 The phases of the Moon

We can only see the part of the Moon that is lit up by the Sun. As the Moon orbits the Earth, we see different parts of the Moon lit up.

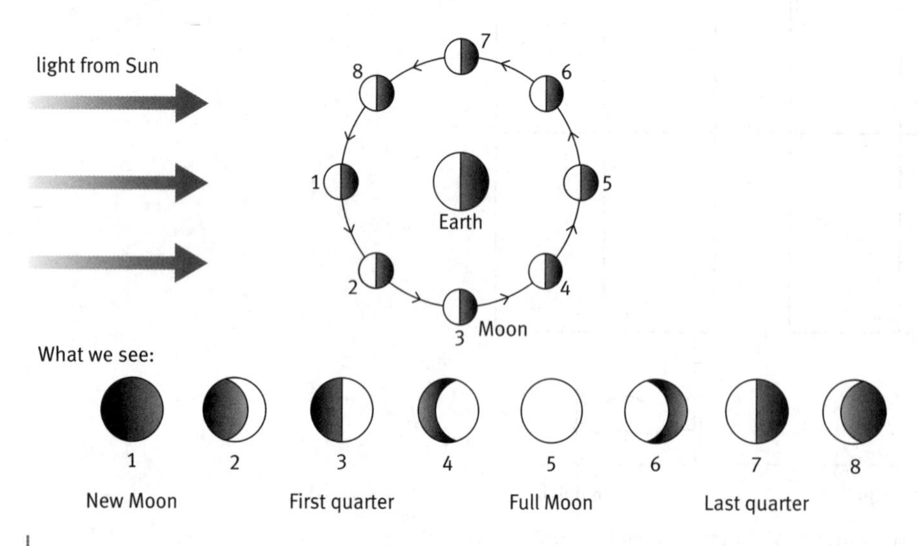

Phases of the Moon.

P7.1.6 What causes an eclipse?

Eclipses involve the Sun and the Moon.

- In a **solar eclipse**, the Moon blocks the Sun's light.
- In a **lunar eclipse**, the Moon moves into the Earth's shadow.

The Moon and Earth both have shadows. The shadows have a region of total darkness (the **umbra**) and a region of partial darkness (the **penumbra**). The Earth's shadow is bigger than the Moon's shadow.

- Where the Moon's umbra touches the surface of the Earth, there is a solar eclipse from inside the area of the umbra. There is a partial eclipse inside the area of the penumbra.
- When the Moon passes into the Earth's umbra, there is a lunar eclipse. Lunar eclipses are more common than solar eclipses because the Earth's shadow is bigger than the Moon's shadow. During a lunar eclipse you can still see the Moon, but it looks red. This is because red light from the Sun is refracted by the Earth's atmosphere, so it can still reach the Moon's surface.

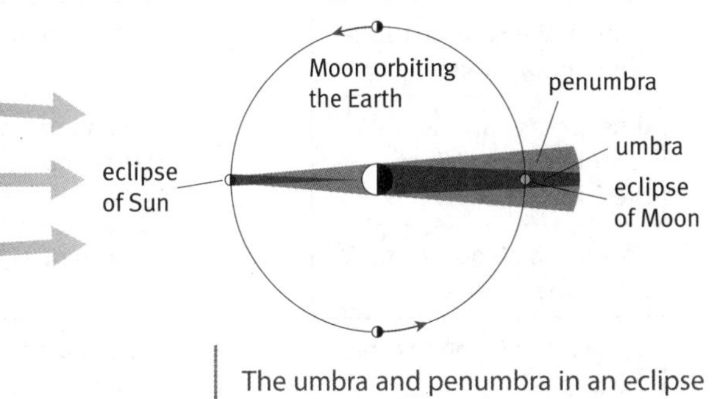

The umbra and penumbra in an eclipse of the Sun and an eclipse of the Moon.

Eclipses are rare because the Moon does not often line up exactly with the Sun. This is because the Moon's orbit is tilted by 5° relative to the plane of the Earth's orbit.

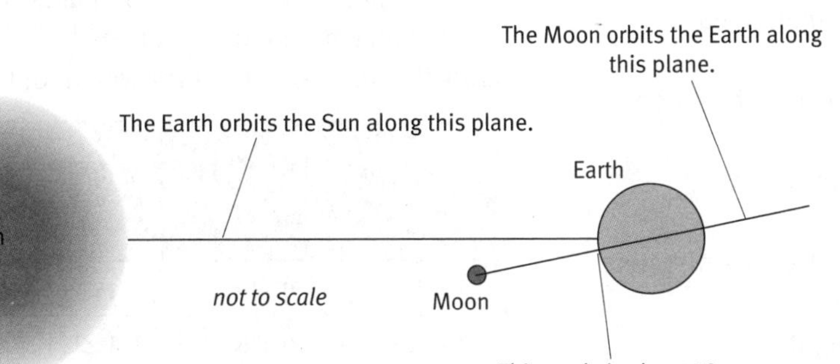

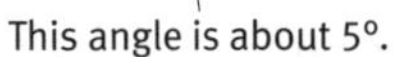

The orbit of the Moon is tilted relative to the plane of the Earth's orbit around the Sun. The effect is exaggerated here.

P7.1.9 Locating objects in the sky

You can describe the stars as though they were lights on the inside of a spinning sphere. This is the **celestial sphere**. The celestial sphere has:

- an axis from the Pole Star through the axis of the Earth
- a celestial equator, which is an extension of the Earth's equator.

Astronomers use two angles to describe the positions of astronomical objects. The angles are measured from a reference point in the sky. The system works wherever you are and whatever the time of day or year.

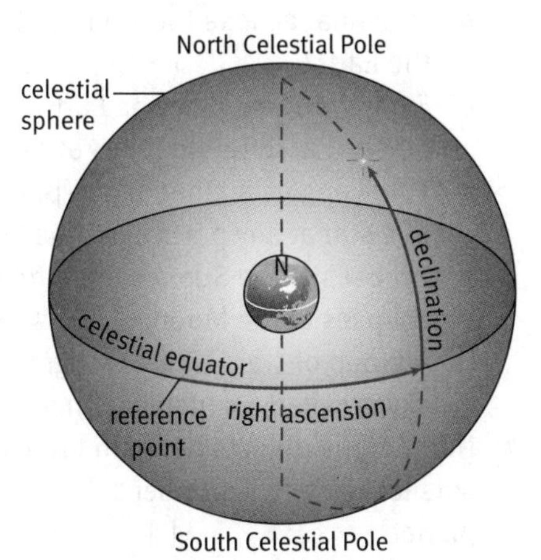

Right ascension measures the angle east from the reference point. Declination measures the angle of the star above or below the celestial equator.

P7.1: Naked-eye astronomy

Use extra paper to answer these questions if you need to.

1 Highlight the one correct word in each pair of **bold** words.

a The **planets / stars** orbit the Sun.

b The Sun appears to move across the sky from **east / west** to **east / west**.

c The Sun appears to move across the sky once every **12 / 24** hours.

d In a **solar / lunar** eclipse, the Moon blocks light from the Sun.

2 Draw a line to match each description to the correct time period.

Description
time for the Earth to rotate once about its axis
time for the Earth to complete one orbit of the Sun
time for the Moon to move across the sky once
time for the Moon to orbit the Earth
time for the Sun to next reappear in the same place in the sky as it is now

Time period
about 28 days
24 hours and 49 minutes
365¼ days
23 hours and 56 minutes
24 hours

3 Write **T** next to the statements below that are **true**. Write corrected versions of the statements that are **false**.

a During a solar eclipse the Earth comes between the Moon and the Sun.

b Sometimes some planets appear to move backwards relative to the stars.

c During the night, stars in the northern hemisphere move in circles about the Pole Star.

d The Moon can only be seen at night.

e In a lunar eclipse the Moon's shadow falls on the Earth.

f From the same place on Earth, different stars can be seen at different times of the year.

g Three angles are needed to pinpoint the position of a star at any particular time.

h Eclipses of the Sun are more frequent than eclipses of the Moon.

i A group of stars with a recognisable pattern in the sky is called a constellation.

4 Number the objects below in order of their size. The smallest object is number 1.

planet [ ]
Solar System [ ]
Universe [ ]
Earth's Moon [ ]
galaxy [ ]
Sun [ ]

5 **Table A** gives data about the Moon during the month of May 2012 for the city of Ulaanbaatar in Mongolia. Use data from Table A to answer parts **a** to **d** below the table.

Date	Time the Moon rises	Time the Moon sets	% of the Moon that is illuminated by the Sun
15 May	02:25	14:50	29
16 May	02:47	15:53	21
17 May	03:09	16:56	13
18 May	03:33	17:59	7
19 May	04:00	19:00	3
20 May	04:30	20:00	1
21 May	05:06	20:56	0
22 May	05:49	21:47	1
23 May	06:38	22:33	5

Table A Ulaanbaatar.

a i Describe the pattern in moonrise times.

ii Explain this pattern.

b Give the date of the new Moon.

c Estimate the date of the full Moon before this new Moon. Explain why you chose this date.

d Draw what the Moon looks like on 23 May.

e Bolormaa looks for the Moon at 22:00 on 22 May. Explain why she does not see it.

Table B gives data about the Moon for London, UK.

Date	Time the Moon rises	Time the Moon sets	% of the Moon that is illuminated by the Sun
15 May	02:39	15:19	27
16 May	02:58	16:25	18
17 May	03:18	17:30	11
18 May	03:40	18:35	6
19 May	04:05	19:39	2
20 May	04:35	20:39	0
21 May	05:11	21:35	0
22 May	05:54	22:25	2
23 May	06:46	23:08	6

Table B London.

f i Describe the difference in moonrise times for Ulaanbaatar and London.

ii Explain this difference.

g Sophie looks for the Moon in London at 22:00 on 22 May. She does not see it. Suggest why not.

1 Charlotte has been observing the December night sky. She noticed a particular group of stars that made a pattern – her star book called this pattern 'Orion'.

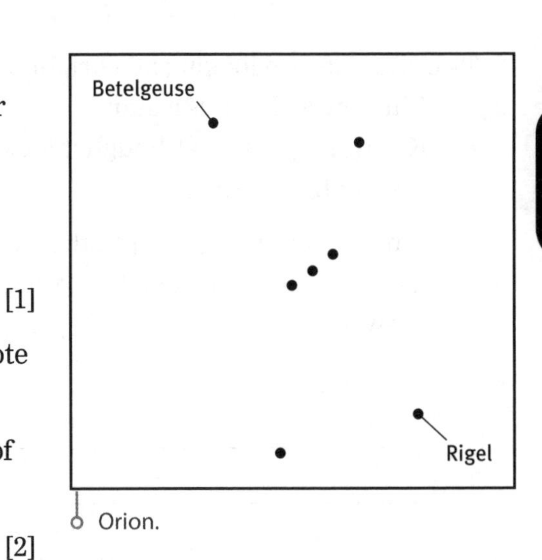

Orion.

a What word is used to describe a pattern of stars like Orion?

_______________________ [1]

b Charlotte measures some angles and makes a careful note of the position of Orion.

Name the two angles she needs to identify the position of a star.

_______________________ and _______________________ [2]

c Charlotte looks in the same position two hours later and finds that Orion has moved. Explain why this is.

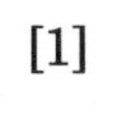

____ [1]

d One star, not in Orion, does not appear to have moved. Explain why it is in the same position.

____

____ [1]

e Charlotte looks again for Orion in June, but can't find it. Explain why this is.

____

____

____ [2]

f Charlotte also notes the time at which the Moon rises for several days. Here are her results.

Date	Time of moonrise
December 24th	16:04
December 25th	17:30
December 26th	18:56
December 27th	20:20

Explain why the moonrise is getting later every day.

____ [1]

Total [8]

2 Two students are looking at a bright object in the night sky.
- Chloe says that it is a star.
- Kai-Wei says that, although it looks like the other stars, it is actually a planet.

a What observation(s) might the students make with the naked eye to help them decide who is right? Explain your answer.

____

____

____

______________________________ [3]

b Identity one difference between a star and a planet.

____

______________________________ [2]

Total [5]

Going for the highest grades

3 Nasir al-Din al-Tusi was an astronomer from Persia. He lived from 1201 to 1274. Diagram 1 is a copy of a diagram drawn by Nasir al-Din al-Tusi to explain how an eclipse of the Sun occurs.

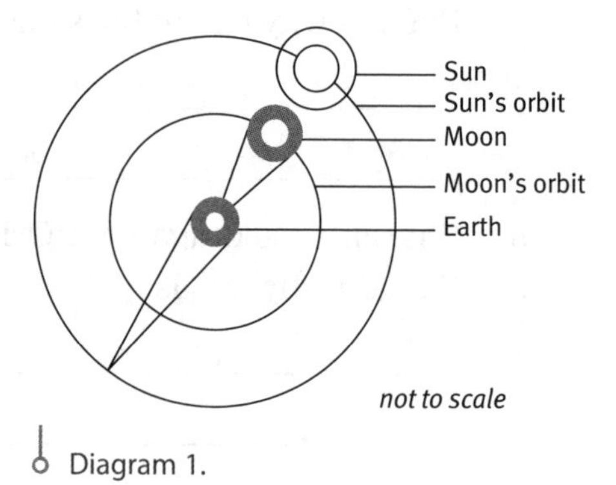

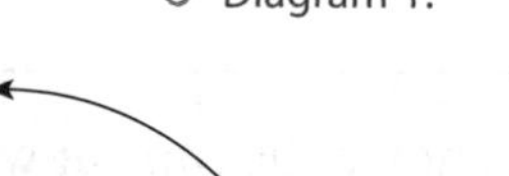
Diagram 1.

Diagram 2 is a modern diagram showing how an eclipse of the Sun occurs.

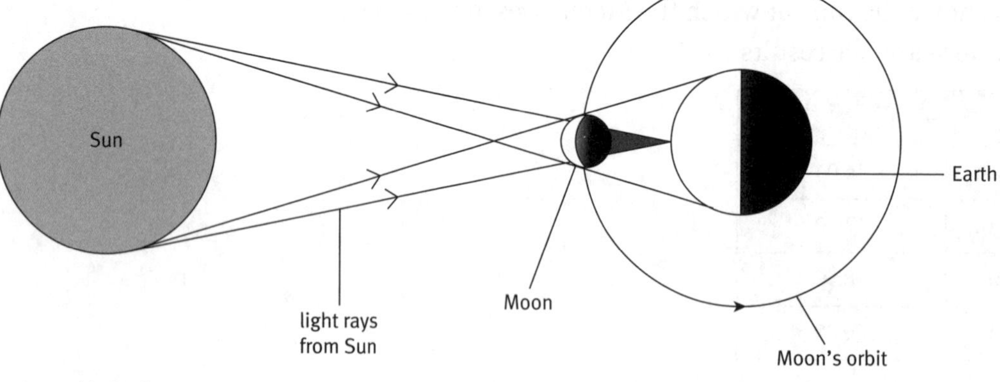

Diagram 2.

Evaluate the two diagrams.

The quality of written communication will be assessed in your answer to this question.

Write your answer on separate paper or in your exercise book.

Total [6]

P7.2 Telescopes and images workout

1 Tick the boxes that show correct ray diagrams.

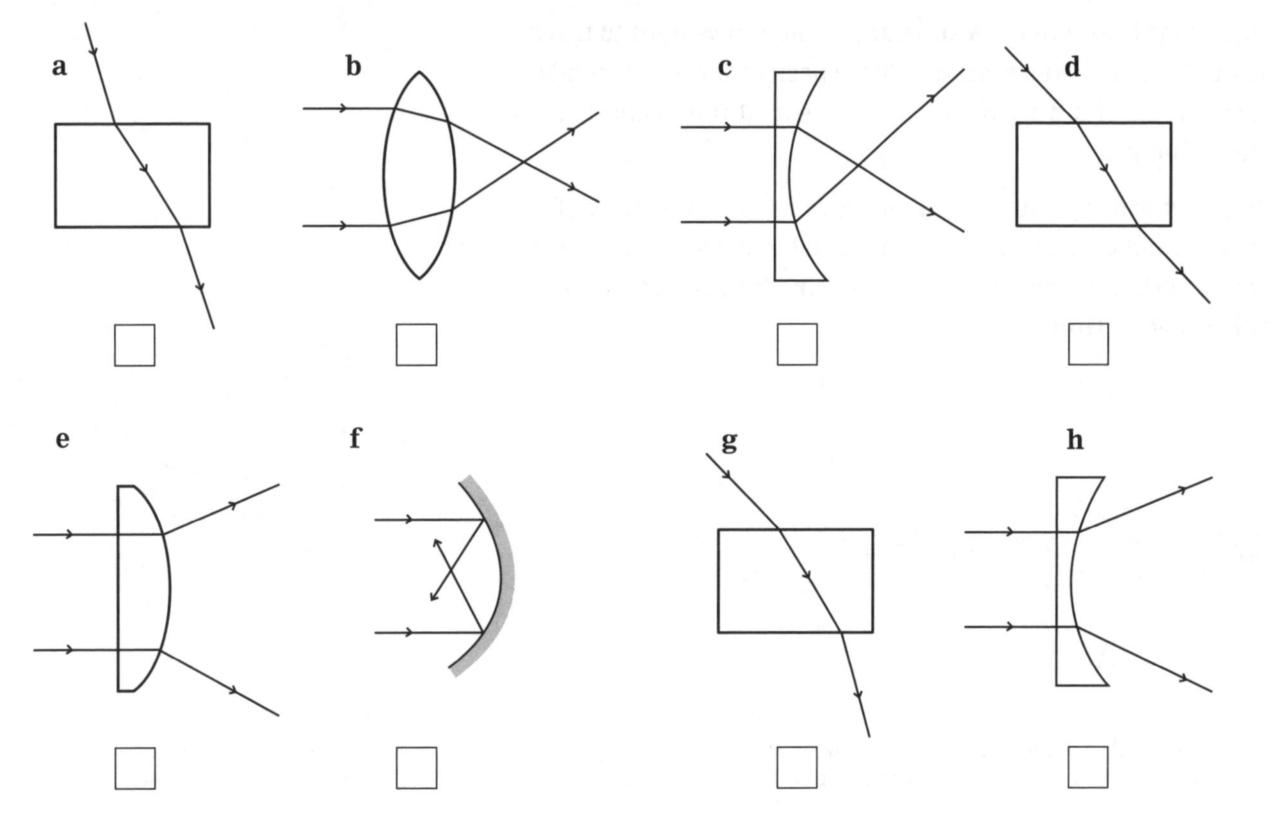

2 All these lenses are made of the same material. Arrange them in order of power – least powerful 1, most powerful 3.

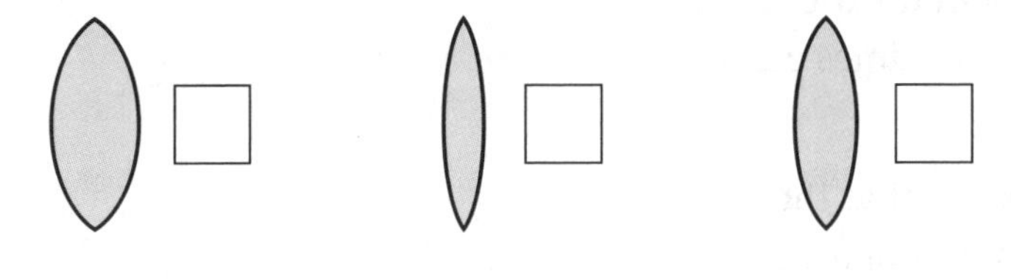

H 3 a Complete this diagram for rays going through a convex lens.

Label the position of the image.

b Draw a (ring) around the correct bold words.

The image in the diagram above is:

- **inverted / right way up**
- **smaller / magnified**

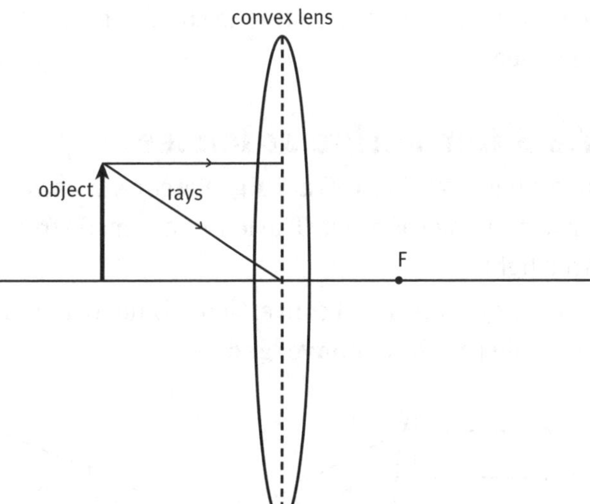

P7.2.1–3 Waves and refraction

Light travels as **waves**. A substance that allows light to travel through it is called a **medium**. The speed of a wave depends on the medium. If a wave travels from one medium to another, its speed changes.

Once a vibrating source has made a wave, the frequency of the wave cannot change. So when the speed of a wave changes, its wavelength also changes. The wave may then change direction. This is **refraction**.

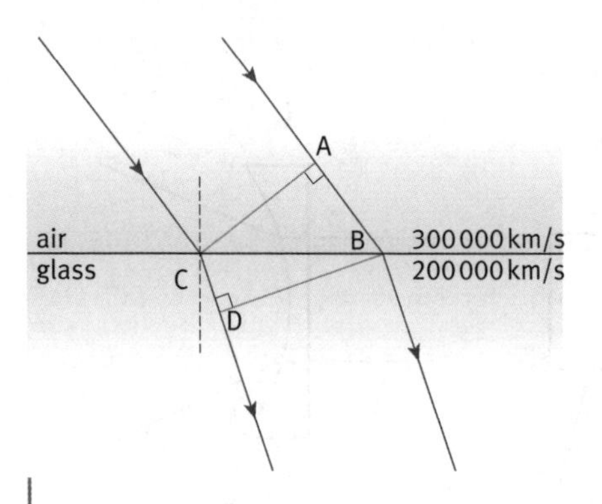

Light is refracted because its speed is greater in air than in glass. Light travels the distance AB in air in the same time as it travels the distance CD in glass.

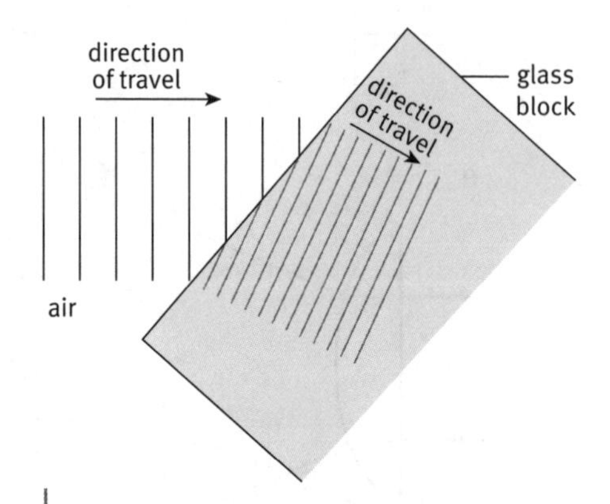

When a wave passes from a less dense medium (air) to a more dense medium (glass) its speed and wavelength change.

Why do we use telescopes?

Ruth is looking at the Moon with her naked eye. Shona is using a telescope to look at the Moon. The telescope has a **magnification** of 50 – it makes the Moon look 50 times bigger and 50 times closer. Shona can see craters that Ruth cannot see.

Now they look at the stars. The stars are so far away that they are points of light, even through the telescope. But Shona's telescope makes the angle between the stars 50 times greater. It also collects more light, so Shona can see dimmer stars that Ruth cannot see.

P7.2.5 Refraction at lenses

Shona's telescope is a **refracting** telescope. It uses **convex lenses** to form an image of a star. The lenses are made from glass. They refract light.

Parallel rays entering a convex lens come to a point called a **focus**. The rays have **converged**.

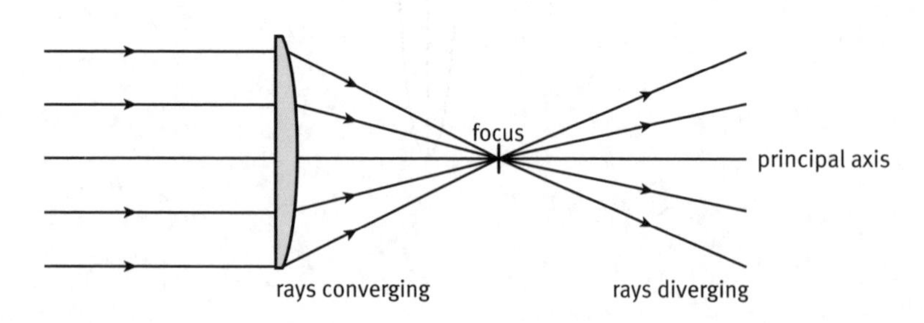

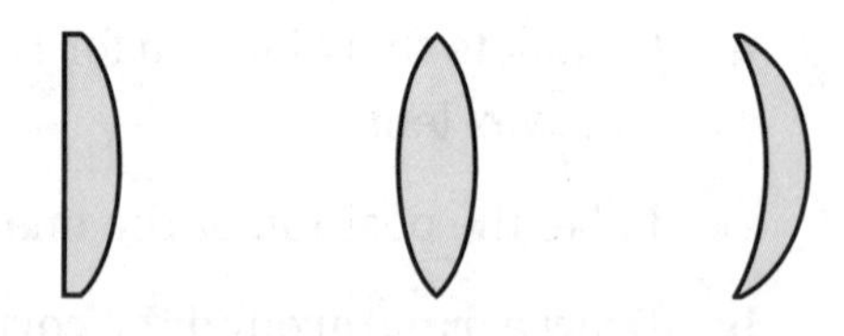

Convex lenses are thicker in the middle than at the edges.

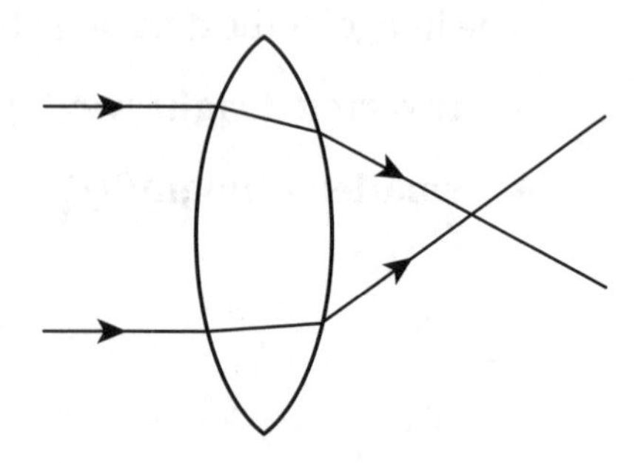

Light is refracted as it enters and leaves a convex lens.

P7.2.4, P7.2.6, P7.2.9 Images in lenses

Use these rules to help you draw ray diagrams:

- Use arrows to show the direction that light is travelling.
- A ray through the centre of a lens does not change direction (ray **a**).
- A ray parallel to the principal axis passes through the focus (ray **b**).
- A ray through the focus emerges parallel to the principal axis (ray **c**).

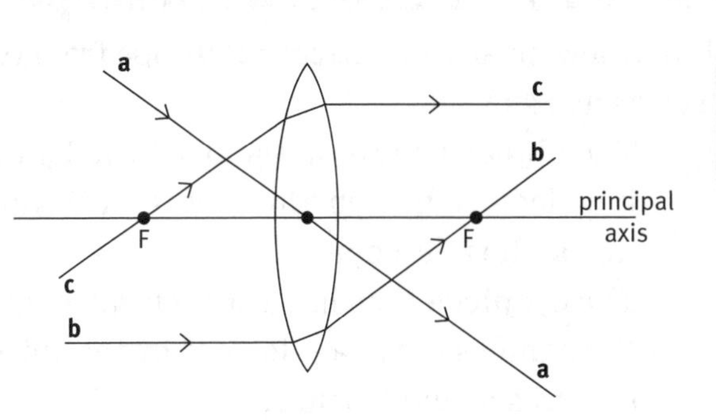

Stars are very far away, so rays reaching Earth from stars are parallel. A convex lens refracts rays from a star through a single point. The point is the **image** of the star – to our eyes, it looks as if the star is at that point.

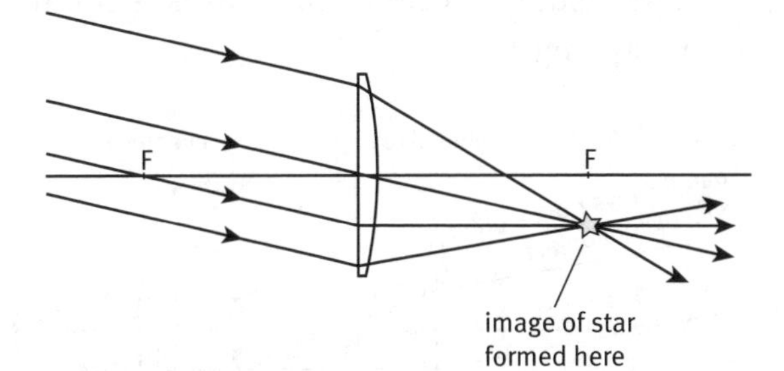

Objects in our own Solar System, such as moons and planets, are closer than stars. Light rays from different parts of the object arrive at a lens at different angles. Rays from the top of the object go to the bottom of the image. The image is upside-down, or **inverted**.

Images of distant objects that are very big, such as galaxies, are also inverted by a convex lens.

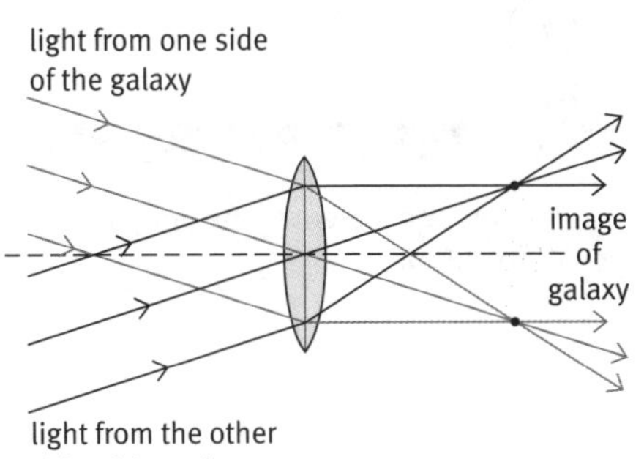

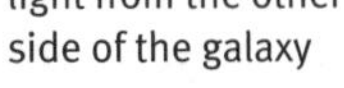

Exam tip

At Foundation tier you need to interpret ray diagrams and draw ray diagrams to show how parallel light comes to a focus. At Higher tier, you also need to draw ray diagrams to show how images of stars, galaxies, moons, and planets in our Solar System are formed.

P7.2.7–9 Focal length and lens power

Rajul holds a convex lens in front of the classroom wall. He moves the lens backwards and forwards until he sees a sharp image of the view outside on the classroom wall. The distance from the lens to the wall is the **focal length** of the lens.

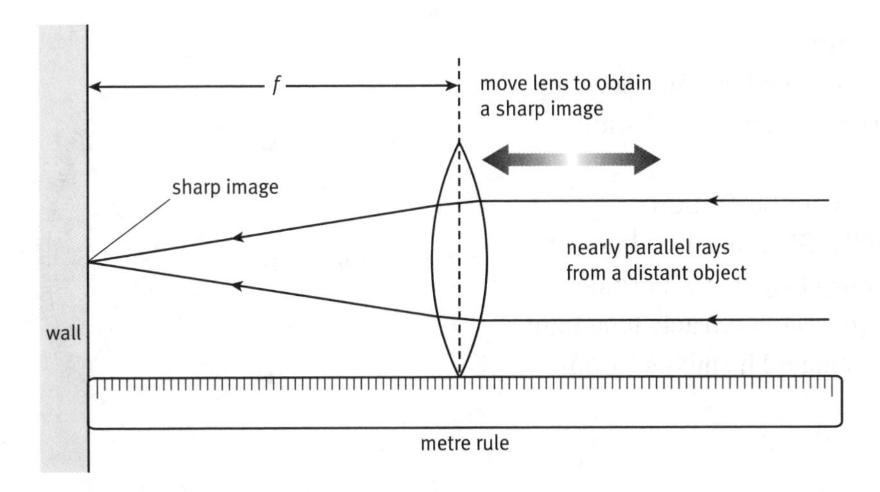

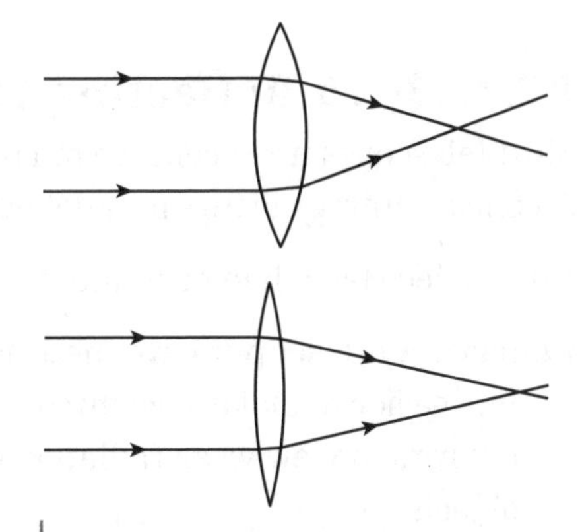

A fat convex lens has a shorter focal length than a thin lens made of the same material. The fat lens refracts light more. It is more powerful.

The **power** of a lens is measured in **dioptres**.

$$\text{power (dioptres)} = \frac{1}{\text{focal length (metres}^{-1})}$$

P7.2.10–11 Inside a telescope

Matthew makes an **optical telescope** from two convex lenses. He looks at a star.

- The **objective lens** has a long focal length (low power). It collects light from the star. It forms an image of the star inside the telescope.
- The **eyepiece lens** has a short focal length (high power). It magnifies the image formed by the objective lens. Matthew sees this magnified image.
- The distance between the lenses is equal to the sum of the two focal lengths.

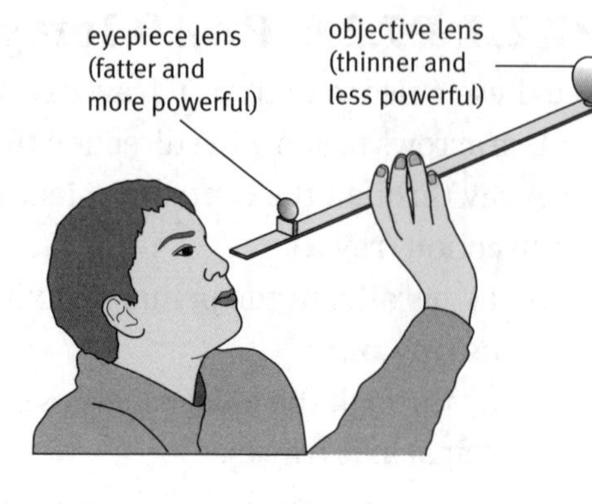

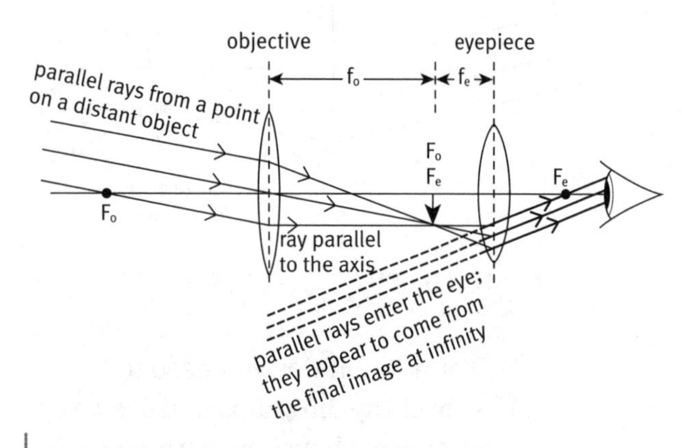

Ray diagram for a refracting telescope.

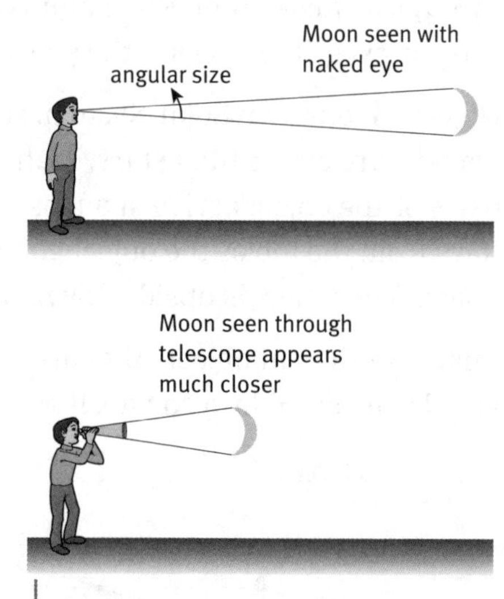

A telescope increases the angular size of the Moon.

P7.2.12 Magnification

H Matthew's telescope makes the angles between stars look much bigger than if he looked with his eyes only. This is the **angular magnification** of the telescope.

$$\textbf{magnification} = \frac{\textbf{focal length of objective lens}}{\textbf{focal length of eyepiece lens}}$$

P7.2.13–15 Reflecting telescopes

Most telescopes use a concave mirror, not a lens, as the objective. A concave mirror brings parallel light to a focus.

An eyepiece lens then magnifies the image from the mirror.

Reflecting telescopes have these advantages:

- It is easier to make a big mirror than a big lens. You need big mirrors to view weak radiation from faint or very distant objects.
- It is hard to make a glass lens with no imperfections.
- A big convex lens is fat in the middle. Glass absorbs light on its way through the lens, so faint objects look even fainter.
- Mirrors reflect all colours the same. A lens refracts blue light more than red, which distorts the image. The next section explains why.

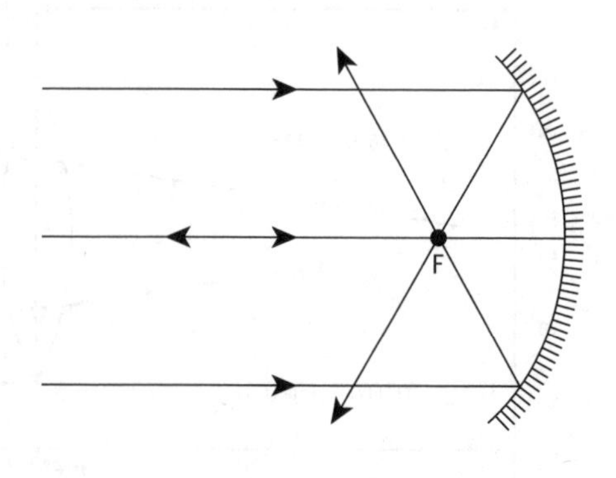

P7.2.20–21 Dispersion

White light is made up of a mixture of colours. Violet light has a higher frequency than red light, and a shorter wavelength. Violet light slows down more in glass, and is refracted more.

In lenses and prisms, refraction splits white light into its colours. This is **dispersion**.

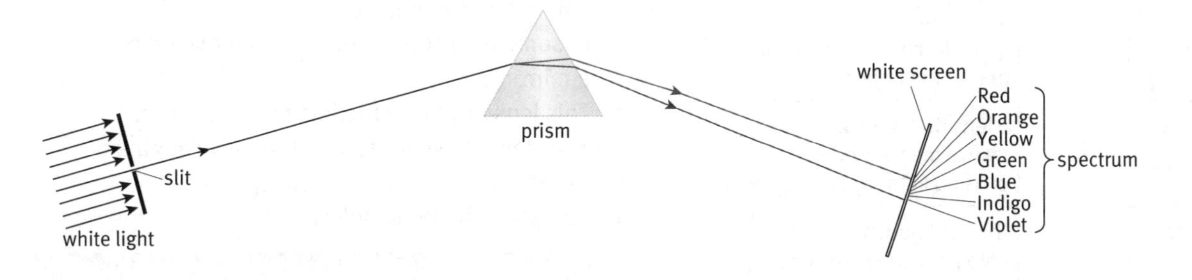

Dispersion in a prism.

Dispersion also occurs at a **diffraction grating** (narrow parallel lines on a sheet of glass). When white light shines on the grating, different colours emerge at different angles. This forms spectra.

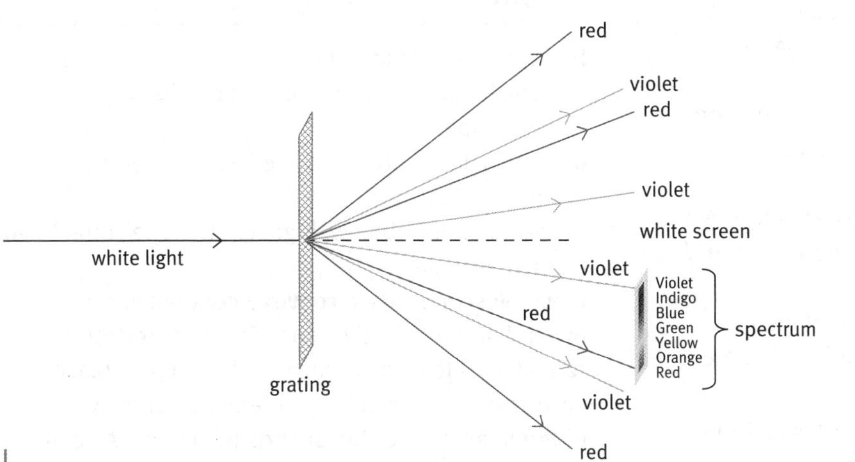

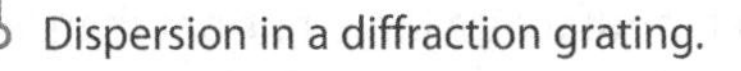

Dispersion in a diffraction grating.

Dispersion distorts images from lenses. This is a problem in telescopes.

Astronomers view stars through **spectrometers** containing prisms or gratings. These show the frequencies of light emitted by the star.

P7.2.16–19 Diffraction

When waves go through a gap, they bend and spread out. This is **diffraction**. The effect of diffraction is greatest when the size of the gap is similar to – or smaller than – the wavelength of the waves.

The light-gathering area of a telescope's objective lens or mirror is its **aperture**. If diffraction occurs at the aperture the image will be blurred. Optical telescopes have apertures much bigger than the wavelength of light to reduce diffraction and form sharp images.

Radio waves have long wavelengths. A telescope that detects radio waves from distant objects needs a very big aperture.

Waves spread out as they pass through the aperture.

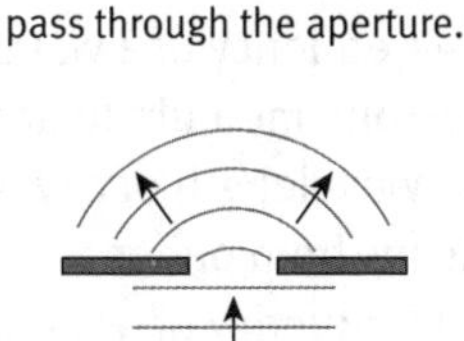

A narrower aperture has more effect.

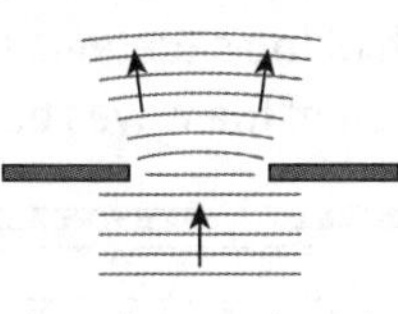

A smaller wavelength gives less diffraction.

Diffraction is greatest when the gap is similar to the wavelength of the waves.

Use extra paper to answer these questions if you need to.

1 Draw lines to match each word to its description.

Word
dioptre
convex lens
spectrum
magnification
focal length
objective lens

Description
the distance between the focus and the centre of a lens
how much bigger an image is than the object
in a telescope, the lens that is nearer the object
a lens that is thicker in the centre than the edges, causing light rays to converge
the continuous band of colours seen when light passes through a prism
the unit for measuring the power of a lens

2 In each sentence below, highlight the one **bold** word that is correct.

a The bouncing back of light at the boundary between two materials is called **refraction / reflection / diffraction.**

b The change of direction of light as it passes from one material to another is called **refraction / reflection / diffraction.**

c The spreading out of a wave as it passes through a small aperture is called **refraction / reflection / diffraction.**

3 Write **T** next to the statements below that are true. Write corrected versions of the statements that are false.

a Light from distant stars reaches Earth as parallel sets of rays.

b When light waves travel from air to glass, they speed up.

c The frequency of a wave changes when it passes from one medium to another.

d The wavelength of a wave stays the same when it passes from one medium to another.

e The frequency of a wave cannot change once it has been made.

f The speed at which a wave travels depends on the medium it is travelling in.

4 In which of the situations shown in the diagrams will the diffraction of the waves be least? Explain your answer.

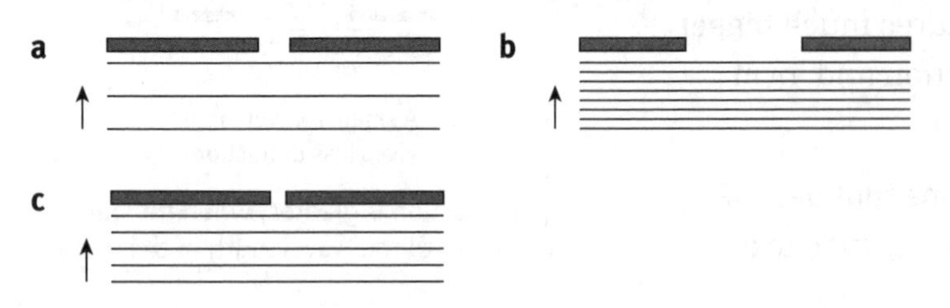

5 Write **O** next to the statements below that are true of an objective telescope lens. Write **E** next to the statements that are true of an eyepiece lens. Write **B** next to the statements that are true of both.

a This lens collects light from the object being observed.

b This lens produces a magnified image of another image.

c This is a converging lens.

d In some telescopes, this lens is replaced by a mirror.

e This lens produces the image that we see.

6 Four students have measured the focal lengths of convex lenses, made of the same glass.

a Complete the table below.

Student	Focal length (cm)	Power (dioptres)
Guy	50	
Kevin	20	
Nikhita	10	
Clare	40	

b Whose lens is thinnest?

c Which pair of lenses would make the best telescope?

d Whose lens should be used as the eyepiece lens in the telescope?

7 In each sentence below, highlight the one correct **bold** word in each pair.

Most telescopes use a **convex / concave** mirror instead of the objective lens. These are **reflecting / refracting** telescopes. One disadvantage of **lenses / mirrors** is that they refract different colours by different amounts. This distorts the image. Also, it is **harder / easier** to manufacture large mirrors than lenses.

8 Calculate the angular magnification for each of the telescopes in the table.

Focal length of objective lens (cm)	Focal length of eyepiece lens (cm)	Magnification
20	5	
30	4	
25	3	

1 This question is about telescopes.

Rebekah makes a telescope.

The ray diagram below shows the arrangement of the lenses forming her telescope. The diagram is not drawn to scale.

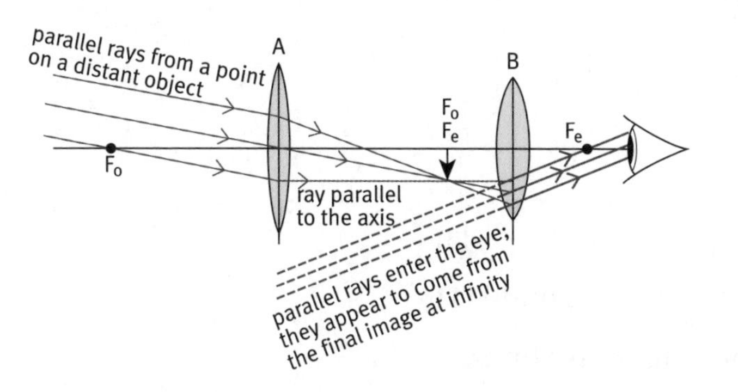

Rebekah's telescope.

Lens A has a focal length of 80 cm. Lens B has a focal length of 5 cm.

a Complete the diagram by labelling:
- the objective lens
- the principal axis
- the focal length of lens A. [3]

b The light rays incident on the telescope are effectively parallel.

Explain why this is.

____

____ [1]

c Calculate the power of lens A and give the unit. Show your working.

Power of lens A = ________ ________ [2]

d Calculate the angular magnification of the telescope. Show your working.

Magnification = ______________ [2]

e Alex makes a telescope.

It has an angular magnification of 20.

Its eyepiece lens has a focal length of 2 cm.

Normally, the distance between the lenses in a telescope is the sum of the focal length of the objective lens and the focal length of the eyepiece lens.

Whose telescope is longer – Rebekah's or Alex's?

Show how you work out your answer.

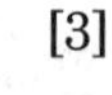

[3]

Total [11]

2 a The diagrams below show waves approaching apertures.

A ☐ B ☐ C ☐

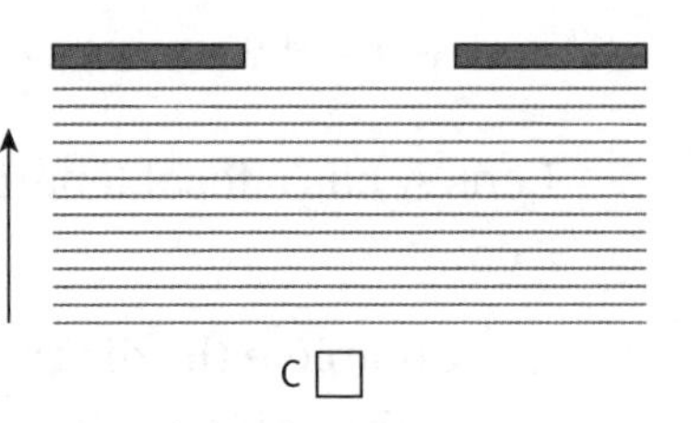

i Tick the box below the diagram in which the diffraction of the waves will be greatest. You might need to use a ruler to help you decide. [1]

ii Explain the choice you made in part (i).

____

____ [1]

H **b** The table shows data about four radio telescopes.

Name and location of telescope	Aperture (m)
Arecibo, Puerto Rico	305
Goldstone, USA	64
Honeysuckle Creek, Australia	26
Lovell, UK	76
Parkes Observatory, Australia	64

i Identify the telescope in which waves of wavelength 3 m would be diffracted least. Explain your choice.

____

____ [2]

ii Astronauts first landed on the Moon in 1969. They sent radio waves to Earth so that people could watch the astronauts on television. The telescopes at Parkes, Honeysuckle Creek, and Goldstone received the radio waves at the same time.

Suggest why the quality of the television pictures from Parkes Observatory was better than the quality of the television pictures from Honeysuckle Creek. The wavelength of the signals sent from the Moon was about 13 cm.

____

____ [2]

Total [6]

P7.3 Mapping the Universe

workout

P 7

1 The graph shows data published by Hubble in 1929. Use these words to complete the graph labels and fill in the gaps below.

accurately away distance of galaxy from Earth distance graph Hubble light measure period redshift speed speed of recession uncertain Universe variable galaxy

Hubble used the _______________ of Cepheid _______________ stars to measure the _______________ to galaxies. The _______________ from these galaxies is redshifted. This means that they are moving _______________, and the _______________ is expanding. The speed of recession can be found from the amount of _______________.

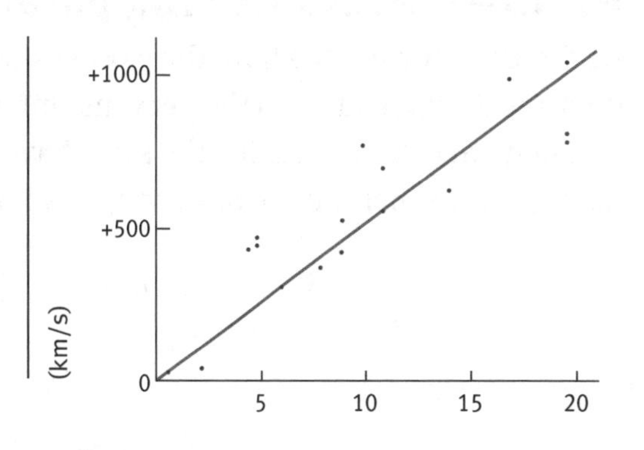

Hubble found that as the distance to a _______________ increases, the _______________ of recession increases. The gradient of the _______________ is called the _______________ constant. Its exact value is _______________ as it is difficult to _______________ the distances _______________.

2 Solve the clues to complete the grid and reveal the name of a famous scientist.

What is she famous for?

1 Our galaxy (5–3)

2 Seen through a telescope, this appears as a fuzzy patch of light (6)

3 Scientist who debated with Curtis about whether there were galaxies outside our own (7)

4 A star that pulses in brightness (7, 8)

5 Distance light travels in a year (5–4)

6 Luminosity depends on the _________ and size of the star (11)

7 Data from an absorption spectrum, used to calculate the recession speed of distant galaxies (8)

P7.3.6 Light-years

A **light-year** is the distance light travels in one year. After the Sun, the nearest stars are about 4 light-years away. This means we see the light that left those stars 4 years ago. Some galaxies are millions of light-years away.

P7.3.1–7, P7.3.17 Using parallax

As the Earth orbits the Sun, the closest stars appear to change their positions relative to the very distant 'fixed stars'. This effect is called **parallax**. Actually, the stars have not moved. It is the Earth, from which we are observing, that has moved.

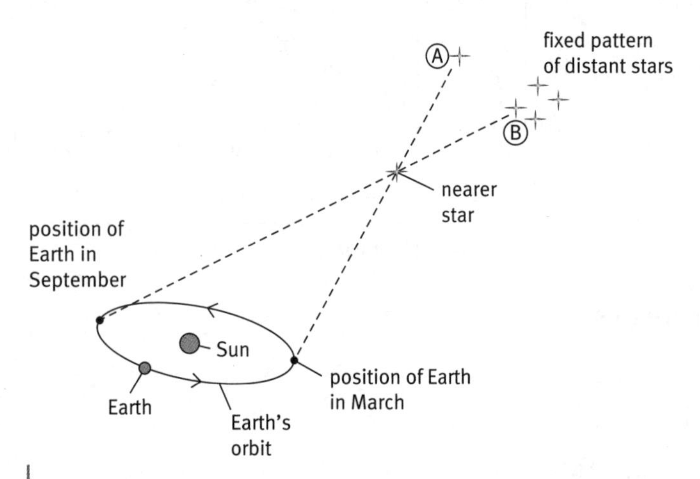

In March the nearer star is in front of star A, but in September it is in front of star B.

The **parallax angle** of a star is half the angle the star has apparently moved in 6 months (when the Earth has travelled from one side of the Sun to the other).

- Parallax angles are tiny. They are measured in **seconds of arc.** (1 second of arc is $\frac{1}{3600}$ of a degree.)
- The smaller the parallax angle, the further away the star:

$$\textbf{distance to star} \text{ (in parsecs)} = \frac{1}{\textbf{parallax angle} \text{ (in seconds of arc)}}$$

A **parsec** (pc) is the distance to a star whose parallax angle is 1 second of arc. Astronomers use parsecs to measure distance.

- A parsec is similar in magnitude (size) to a light-year.
- Distances between stars within a galaxy (interstellar distances) are usually a few parsecs.
- Distances between galaxies are measured in **megaparsecs** (Mpc).

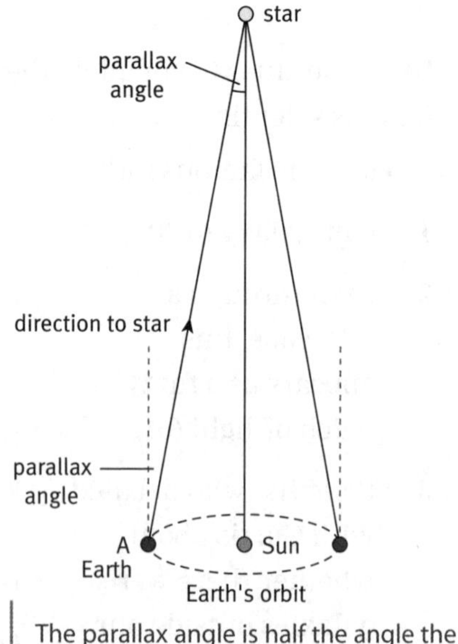

The parallax angle is half the angle the star has apparently moved in 6 months.

P7.3.8–9 Star luminosity

The **luminosity** of a star is the amount of radiation it emits every second. Luminosity depends on temperature and size – the hotter and bigger the star, the more energy it radiates per second.

The **observed intensity** of a star describes the radiation reaching Earth from a star. Observed intensity depends on the luminosity of the star and its distance from Earth. The further away a star is, the less bright it seems to be. This is because the light has spread out over a bigger area.

Astronomers can calculate the distance to a star if they know its luminosity and its observed intensity.

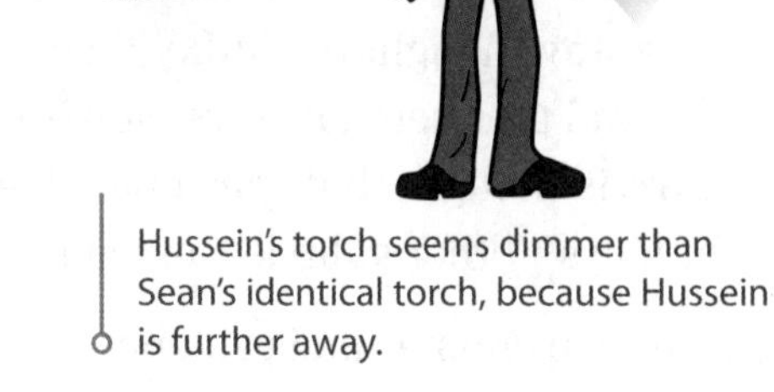

Hussein's torch seems dimmer than Sean's identical torch, because Hussein is further away.

P7.3.10 Cepheid variable stars

A **Cepheid variable star** is a star whose brightness varies. There is a regular pattern of change as the star gets bigger and smaller. The time between peaks of brightness is the **period** of a Cepheid variable star.

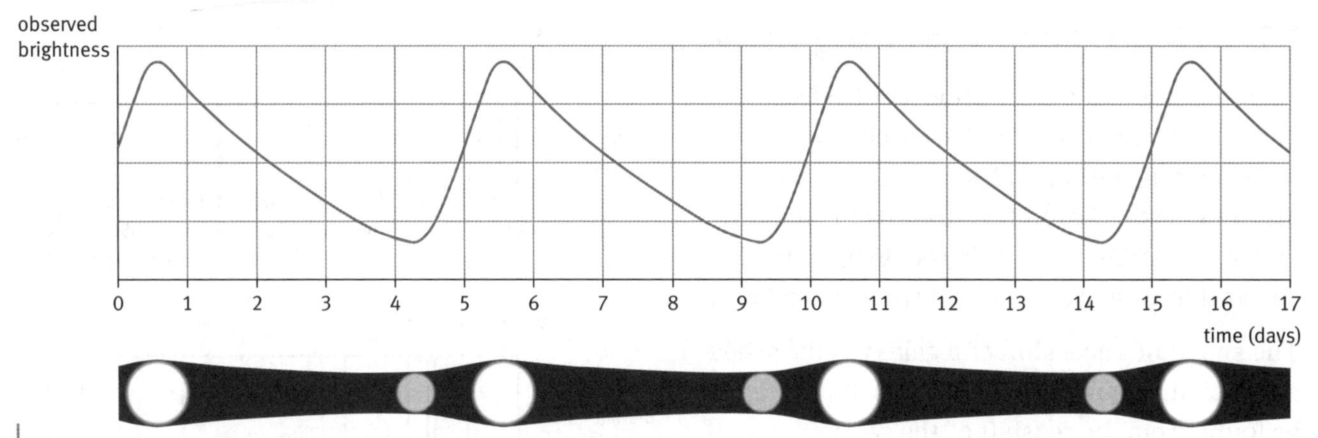

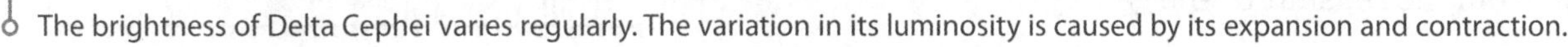

The brightness of Delta Cephei varies regularly. The variation in its luminosity is caused by its expansion and contraction.

P7.3.10–11 Using Cepheid variables to estimate distance

Henrietta Leavitt discovered a correlation between the luminosity of a Cepheid variable star and its period – the greater the luminosity, the longer the period.

Distances to many galaxies are too large to measure using parallax. Instead, astronomers look for a Cepheid variable star within a galaxy. They use the relationship between luminosity and period to work out the distance to the star.

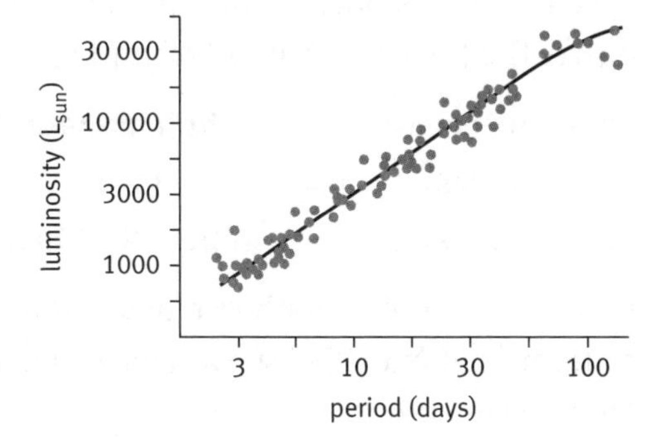

H Astronomers calculate distances to a Cepheid variable star like this:

- Measure the period.
- Use the period to work out the luminosity.
- Measure the observed brightness.
- Compare the observed brightness with the luminosity to work out the distance.

P7.3.12–16 Observing nebulae and galaxies

Telescope observations show that our galaxy, the Milky Way, is made up of millions of stars. The Sun is one of these stars.

In the 1920s, astronomers were puzzled by fuzzy patches of light seen through telescopes. They called them **nebulae**. Nebulae have different shapes, including spirals. Astronomers debated spiral nebulae.

- **Shapley** thought the Milky Way was the entire Universe. He said that nebulae were clouds of gas within the Milky Way.
- **Curtis** thought that spiral nebulae were huge, distant clusters of stars – other galaxies outside the Milky Way.

Neither astronomer had evidence strong enough to settle the argument. Later, **Hubble** found a Cepheid variable in a spiral nebula, Andromeda. Hubble measured the distance from Earth to the star. The star was further away than any star in the Milky Way. He concluded that the star was in a separate galaxy.

Cepheid variable stars have been used to show that most spiral nebulae are distant galaxies. There are billions of galaxies in the Universe.

P7.3.18–21 The expanding Universe

Astronomers study absorption spectra from distant galaxies. Compared to spectra from nearby stars, the black absorption lines for distant galaxies are shifted towards the red end of the spectrum. This is **redshift**. Redshift shows that galaxies are moving away from us.

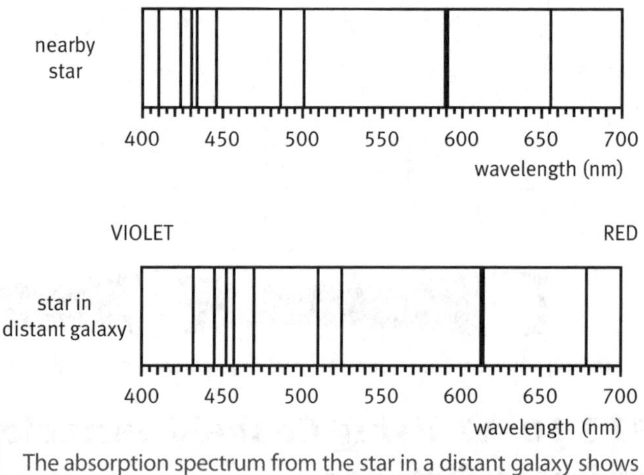

The absorption spectrum from the star in a distant galaxy shows redshift compared to the spectrum from the closer star.

The **speed of recession** of a galaxy is the speed at which it is moving away from us. Its value can be found from the redshift of the galaxy.

Hubble measured the distance to Cepheid variable stars in several galaxies. He found that the further away a galaxy is, the greater its speed of recession. (In this equation, you can use either the top or bottom set of units.)

speed of recession = **Hubble constant** × **distance**

speed of recession	Hubble constant	distance
(km/s)	(s^{-1})	(km)
(km/s)	(km/s per Mpc)	(Mpc)

Other astronomers gathered data from Cepheid variable stars in different galaxies. These data have given better values of the Hubble constant.

The fact that galaxies are moving away from us suggests that the Universe began with a **big bang** about 14 thousand million years ago.

H Distant galaxies seem to be moving away faster than nearby galaxies. This has led scientists to conclude that space itself is expanding.

Exam tip

In Higher-tier exams, you might need to use the Hubble equation to calculate speed of recession, the Hubble constant, or distance. In Foundation-tier exams, you will only be asked to calculate the speed of recession.

Use extra paper to answer these questions if you need to.

1 Draw lines to match each word to its description.

Word
parallax
parsec
light-year
nebula
galaxy
Cepheid variable

Description
a group of thousands of millions of stars
the way that closer stars seem to move over time relative to more distant ones
a star whose brightness varies periodically
the distance that light travels in 1 year
the distance to a star with a parallax angle of 1 second of arc
name once given to any fuzzy object seen in the night sky

2 Use the data in the table to answer the questions below.

Star	Parallax angle (seconds of arc)
A	0.52
B	0.015
C	0.084
D	0.17

a Which star is closest?
b Which star is furthest away?
c Calculate the distance of star D from Earth.

3 Use the words and phrases in the box to fill in the gaps in the sentences below. Each word may be used once, more than once, or not at all.

parallax angle　**size**　**distance from Earth**
luminosity　**observed intensity**　**height**

a The _______ of a star is related to its distance from Earth.
b The period of a Cepheid variable is related to its _______.
c The speed at which a galaxy is moving away is related to its _______.
d The _______ of a star depends on its temperature and size.
e The observed intensity of light from a star depends on its _______ and _______.

4 In the 1920s there was a great debate between the astronomers Curtis and Shapley. Hubble later provided evidence showing that one of them was correct.

Copy and complete the conversation below.
Curtis: I think that nebulae are distant galaxies outside our galaxy.
Shapley: You are wrong. I think they are...
Hubble: I have used a Cepheid variable star to ....
Hubble: My measurements show that...
Hubble: So _____ must be right.

5 a Complete the table.

Star	Parallax angle (seconds of arc)	Distance (parsecs)
Kapteyn's Star	0.250	
Procyon	0.285	
Sirius	0.370	

b Explain why parallax cannot be used to find stellar distances greater than 100 parsecs.
c The star Alpha-Centauri is 1.2 parsecs away from Earth. Which of the following gives its distance in light-years?
A 4.3 light-years
B 43 light-years
C 4300 light-years
D 4.3 light-seconds

6 Write **T** next to the statements that are true. Write corrected versions of the statements that are false.
a The parallax angle of a star is the angle moved against a background of very distant stars in 6 months.
b A parsec is similar in magnitude to a light-year.
c A megaparsec is one billion parsecs.
d Typical interstellar distances are measured in megaparsecs.
e Cepheid variable stars pulse in brightness.
f The smaller the period of a Cepheid variable, the greater its luminosity.
g The Sun is the only star in the Milky Way galaxy.
h Distances between galaxies are measured in megaparsecs.
i The further away a star is, the lower its luminosity.
j Scientists believe the Universe began with a big bang about 14 million years ago.

7 Put these sentences in order to show how an astronomer can work out the distance to a Cepheid variable star.
A Plot the observed brightness of the Cepheid variable over several months.
B Compare the observed intensity with the luminosity to find the distance.
C Use the period to work out its luminosity.
D Find the period of the Cepheid variable from the graph.

8 a Calculate the speed of recession of a galaxy that is 3×10^{21} km away from Earth. Assume the Hubble constant to be $2 \times 10^{-18}\ s^{-1}$.

H **b** Another galaxy has a speed of recession of 2000 km/s. How far away is it?
c The distance to a Cepheid variable star is measured as 4.3×10^{20} km, and the speed of recession of its galaxy is 1990 km/s. What value does this give for the Hubble constant?

1 Look at this graph showing the variation of brightness of the star TU Cassiopeiae.

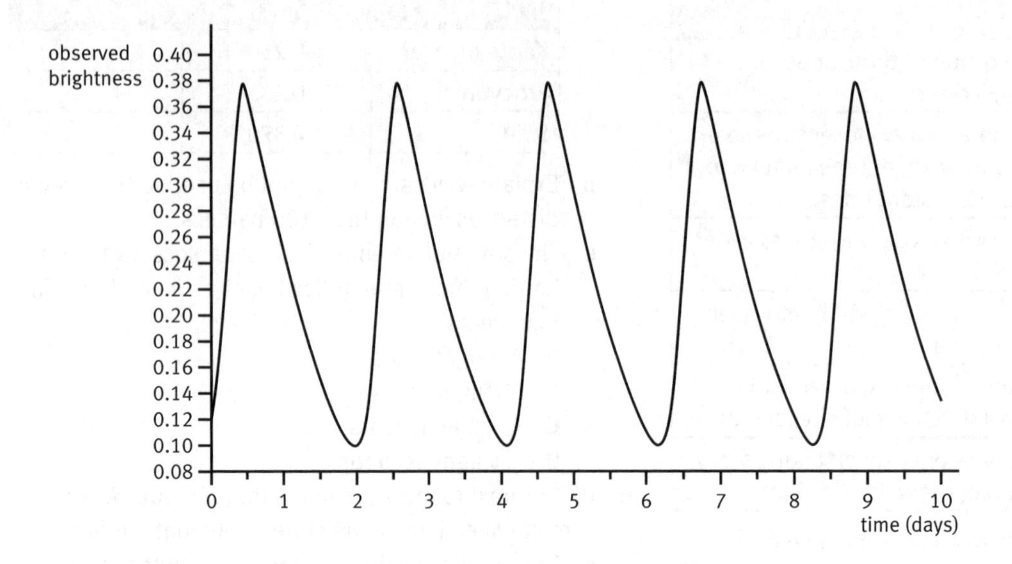

a What sort of star is TU Cassiopeiae? Draw a (ring) around the correct answer.

neutron star **supernova** **Cepheid variable** **red giant**

[1]

b Use the graph above to work out the period of the star.

____ [1]

c The period can be linked to the luminosity of a star.

Use your answer to part **b** and the graph below to work out the luminosity of TU Cassiopeiae.

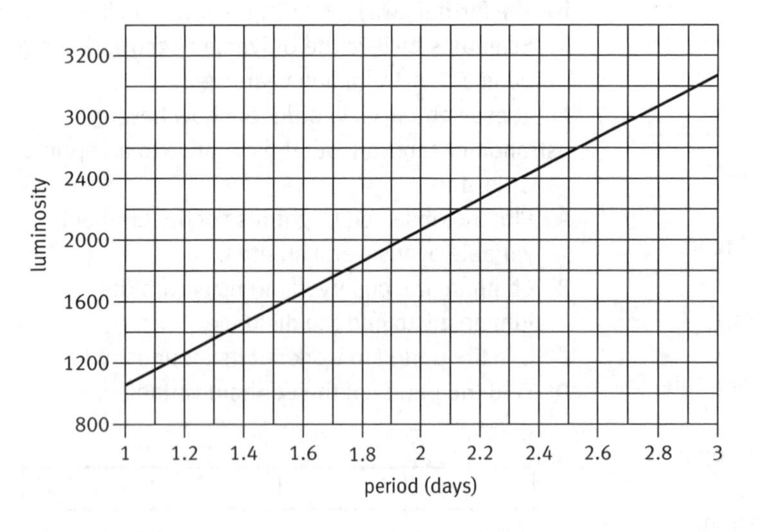

[1]

Answer: ______ units

d The tables give data about three other stars.

Table A

Name of star	Luminosity (relative to the Sun)	Distance from Earth (pc)
Epsilon Cassiopeiae	2500	126
Alpha Cassiopeiae	676	70
Beta Cassiopeiae	27	17

Table B

List of stars in order of observed intensity (highest observed intensity – most bright – first)
Alpha Cassiopeiae
Beta Cassiopeiae
Epsilon Cassiopeiae

Use the data in Table A to explain the data in Table B.

✎ The quality of written communication will be assessed in your answer to this question.

Write your answer on separate paper or in your exercise book. [6]

Total [9]

2 A data book gives parallax angles for various astronomical objects, as shown in the table on the right.

Name	Parallax angle/ seconds of arc
Sirius A	0.38
Orion nebula	0.002
61 Cygnus	0.286
Crab nebula	0.0005

a Explain what is meant by parallax.

____

____

____

[2]

b Without doing any calculation, which object is closest to Earth?

Explain how you know.

____

____

[2]

c Calculate the distance to 61 Cygnus in parsecs. Show your working.

Distance = __________ parsecs [2]

d 61 Cygnus is a pair of stars in the constellation of Cygnus.

A constellation is a group of stars that forms a recognisable pattern from Earth.

Alpha Cygni is another star in the constellation of Cygnus.

Its distance from Earth is approximately 440 pc.

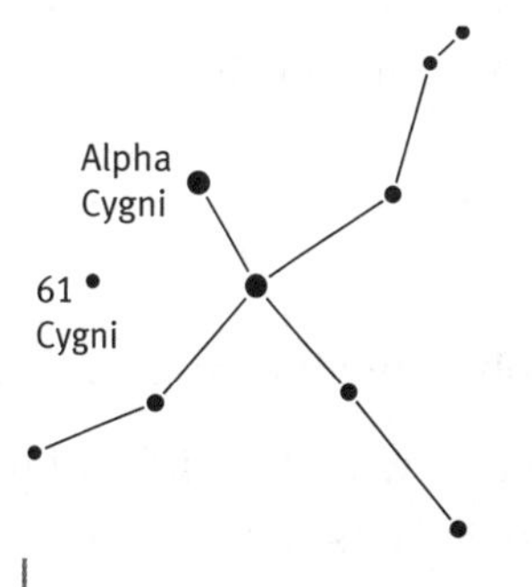

The constellation of Cygnus. The dots represent stars. The lines show the swan-like shape of the constellation.

Compare the distances from Earth of Alpha Cygni and 61 Cygni.

i What conclusion can you make from this data about the stars in a constellation?

___

___ [2]

ii Suggest what extra data you could collect to become more confident in your conclusion.

___ [1]

Total [9]

3 This question is about galaxies.

a Explain what a galaxy is.

___

___ [1]

b What is the name of our galaxy? Draw a ring around the correct answer.

Orion **Andromeda** **Milky Way**

Cygnus **Cassiopeia** [1]

c The distance to a certain star has been measured as 230 megaparsecs. Is the star in our galaxy or in another? Explain how you know.

___

___ [1]

H **d** The astronomer Edwin Hubble found evidence that distant galaxies are all moving away from the Earth. His result can be written as an equation:

speed of recession (km/s) = Hubble constant (s^{-1}) × distance to galaxy (km)

A galaxy estimated to be 1.5×10^{21} km away has a measured speed of recession of 3500 km/s. Use this equation to calculate a value for the Hubble constant. Show your working.

Hubble constant = ___ s^{-1} [2]

e Explain why it is difficult to get a precise value for the Hubble constant.

___

___ [1]

Total [6]

1 This flow chart shows how stars are thought to be formed and change. Fill in the blanks.

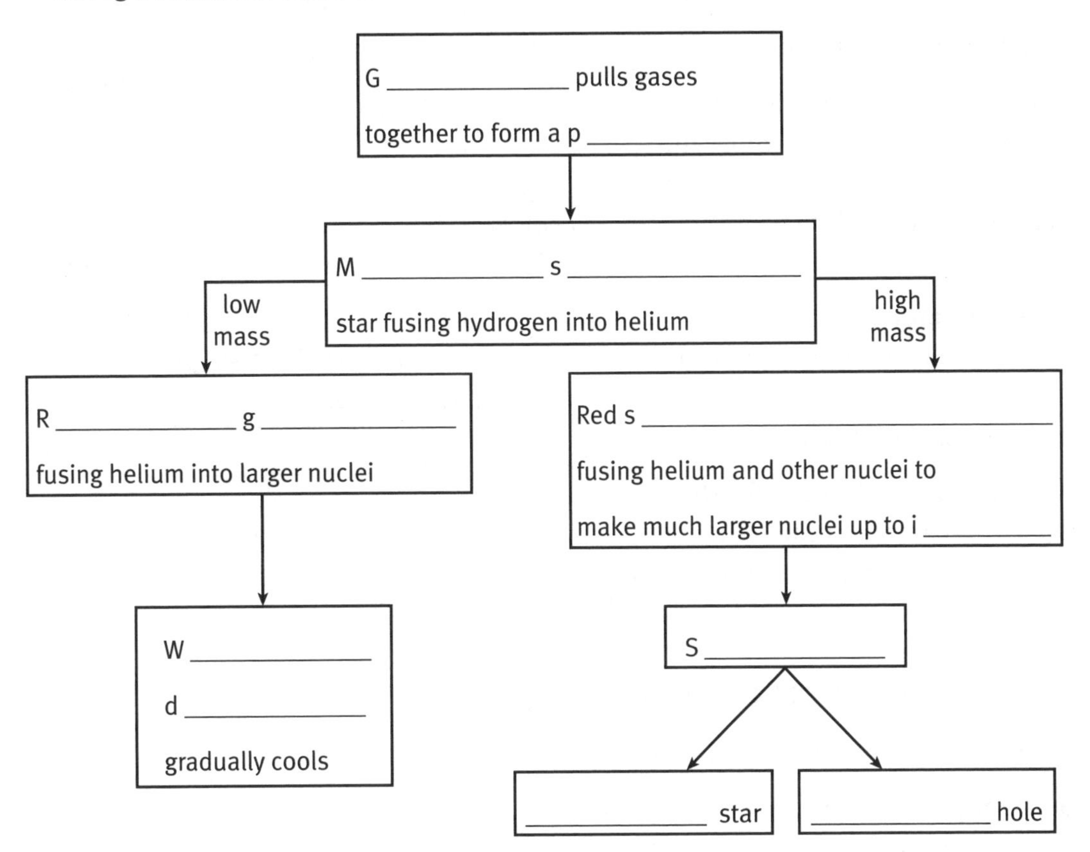

2 Tick the graph(s) showing the correct relationship between the pressure and volume of gas at constant temperature.

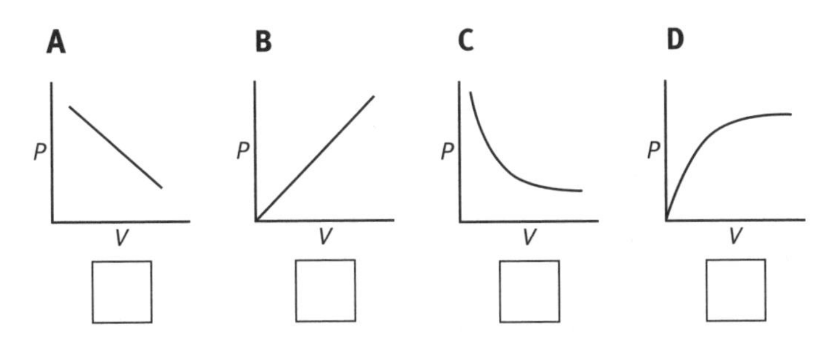

3 Tick the graph(s) showing the correct relationship between the pressure and temperature of a constant volume of gas.

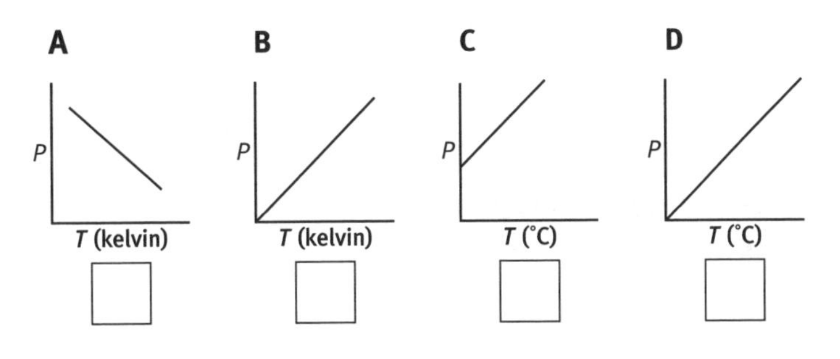

P7.4.1 Star spectra and temperature

Hot objects, including stars, emit energy across all wavelengths of the electromagnetic radiation.

Different stars emit different amounts of radiation at different frequencies, depending on their temperature. This is why the coolest stars appear red, slightly hotter ones look orange, then yellow, then white. The very hottest stars are blue–white.

Astronomers use a **spectrometer** to:

- measure how much radiation is emitted at each frequency
- identify the **peak frequency** of a star.

The peak frequency gives an accurate value for the temperature of a star. The greater the peak frequency, the higher the temperature. The luminosity of a star also increases with temperature.

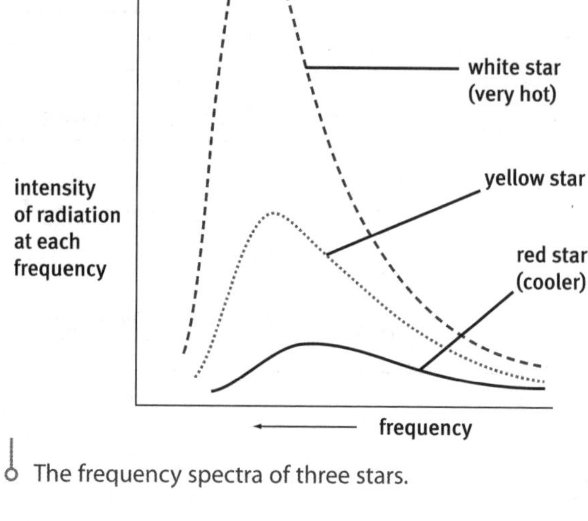

The frequency spectra of three stars.

P7.4.2–4 Identifying elements in stars

Astronomers use spectra from stars to identify their elements.

The surface of the Sun emits white light. As the light travels through the Sun's atmosphere, atoms in this atmosphere absorb light of certain frequencies.

The light that travels on has these frequencies missing. When the light is spread into a spectrum there are dark lines across it. This is the **absorption spectrum** of the Sun.

Each element produces a unique pattern of lines in its absorption spectrum. Astronomers identify the elements in stars by comparing star absorption spectra to those of elements in the lab.

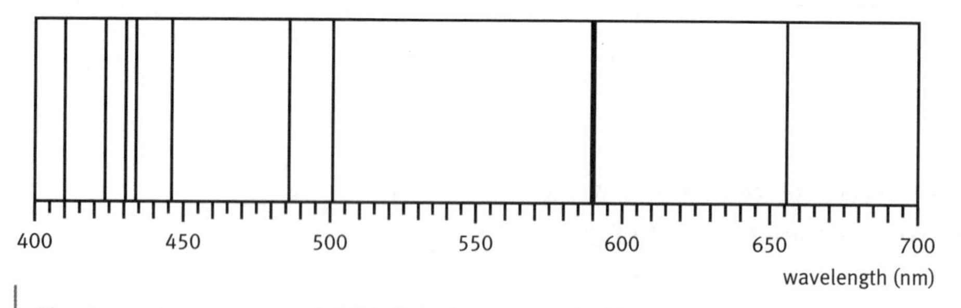

The absorption spectrum of visible light from a star. The black lines show the wavelengths of radiation that are absorbed by atoms of elements in the atmosphere of the star.

H Electrons in an atom of an element can only have certain energy values. The energy levels occupied by electrons are different for every element.

Electrons move between energy levels if they are given exactly the right amount of energy. When light from a star passes through its atmosphere, some of the photons (packets of energy) have energy of exactly the right frequency to move electrons to higher energy levels. It is these frequencies that are missing from absorption spectra.

If electrons are given enough energy they can leave the atom completely. This is **ionisation**.

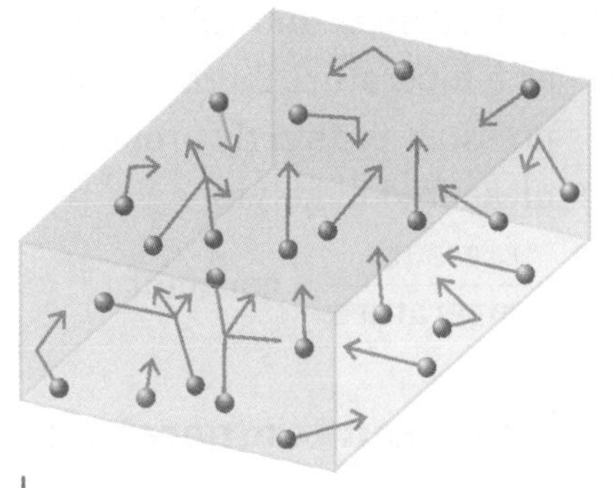

Particles in gases move randomly in all directions.

P7.4.5–10 How do gases behave?

Stars are giant balls of hot gases. To understand stars, you need to understand gases.

The particles of a gas move very quickly in random directions. When they hit the sides of a container they exert a force as they change direction. This causes gas pressure.

Pressure and volume

If you decrease the volume of the container, the particles hit the sides more often. The pressure increases.

Volume and pressure are **inversely proportional**. For a fixed mass of gas at constant temperature:

- as volume increases, pressure ~~increases~~ decreases
- pressure × volume = constant

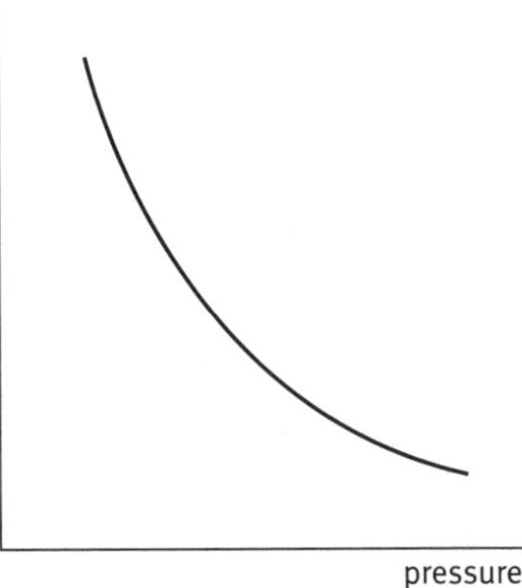

Increasing the pressure on a gas reduces its volume (at constant temperature).

Pressure and temperature

The hotter a gas, the more energy its particles have, and the faster they move. The faster the particles move, the harder and more often they hit the sides of the container.

If you cool a gas, its particles lose energy. They move more and more slowly. At the lowest imaginable temperature the particles would have no energy. They would stop moving, and would never hit the sides of the container. This lowest theoretical temperature is **absolute zero**. It corresponds to –273 °C.

Scientists sometimes use a temperature scale that starts at absolute zero, called the **kelvin scale**. The divisions are called **kelvin (K)** instead of degrees Celsius.

Temperature in K = temperature in °C + 273

Temperature in °C = temperature in K – 273

For a fixed mass and volume of gas:

- as temperature increases, pressure increases
- pressure is directly proportional to the temperature measured in kelvin
- $\frac{\text{pressure}}{\text{temperature}} = \text{constant}$

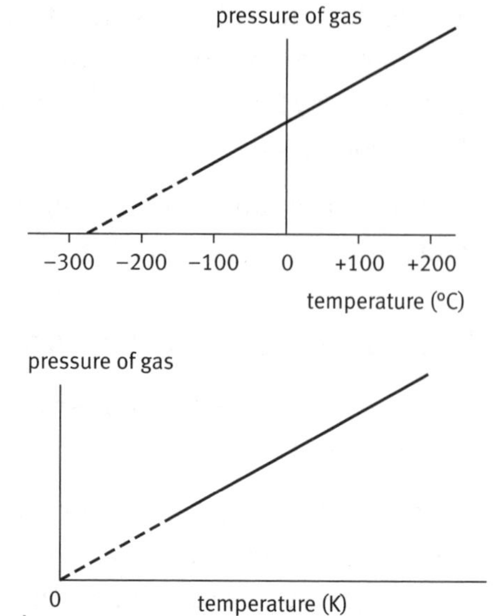

Increasing the temperature of a gas increases its pressure (at constant volume).

Volume and temperature

If you decrease the temperature of a gas at constant pressure, its volume decreases. At a temperature of absolute zero, the volume would theoretically be zero.

For a fixed mass of gas at a fixed pressure:

- as temperature increases, volume increases
- volume is directly proportional to the temperature measured in kelvin
- $\frac{\text{volume}}{\text{temperature}} = \text{constant}$

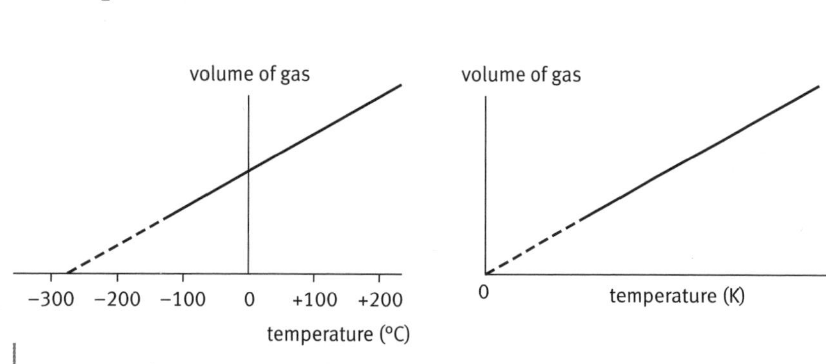

Increasing the temperature of a gas increases its volume (at constant pressure).

P7.4.11–12 What is a protostar?

Gravity compresses a cloud of hydrogen and helium gas. The gas particles get closer and closer. The volume of the gas cloud decreases.

As the gas particles fall towards each other they move faster and faster. Temperature and pressure increase. This mass of gas is called a **protostar**.

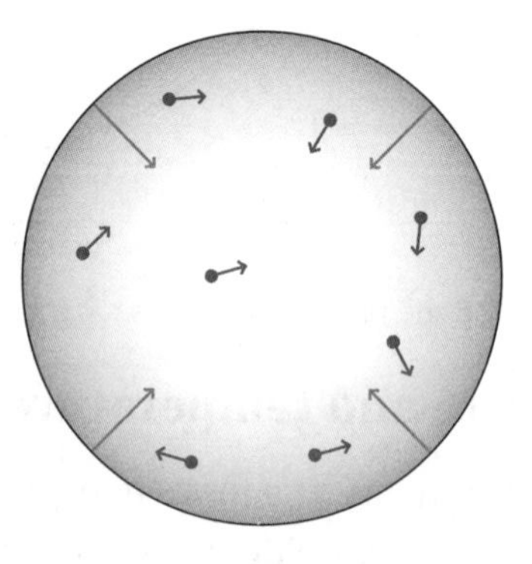

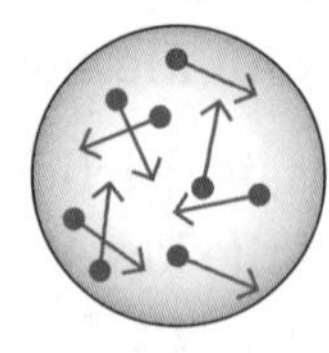

P7.4.13–15, P7.4.17 What is nuclear fusion?

In the early twentieth century scientists discovered nuclear reactions. They thought these reactions might be the source of energy in the Sun.

When hydrogen nuclei come together closely enough, they may join to form nuclei of another element, helium. The process releases energy. This is **nuclear fusion**.

A protostar becomes a star when fusion begins. Nuclear fusion happens in all stars, including the Sun.

The **nuclear equation** below shows one fusion reaction that happens in the Sun.

$${}^{1}_{1}H + {}^{1}_{1}H \rightarrow {}^{2}_{1}H + {}^{0}_{1}e^{+}$$

The product of the reaction above may then fuse with another hydrogen nucleus to form an isotope of helium:

$${}^{1}_{1}H + {}^{2}_{1}H \rightarrow {}^{3}_{2}He$$

In a nuclear equation you must balance:

- mass (top numbers)
- charge (lower numbers)

The symbol ${}^{0}_{1}e^{+}$ represents a **positron**. Positrons are emitted in some nuclear reactions to conserve charge.

In fusion reactions, the total mass of product particles is slightly less than the total mass of reactant particles. The mass that is lost has been released as energy.

Exam tip

In the exam you may be asked to complete and interpret nuclear equations. Practise!

H You can use **Einstein's equation** to calculate the energy released in nuclear fusion and fission reactions:

$$E = mc^2$$

energy released = mass lost × (speed of light in a vacuum)^2

P7.4.16, P7.4.19–21, P7.4.23–28 Inside stars

Main-sequence stars

Stars like the Sun, which fuse hydrogen to form helium, are **main-sequence stars**.

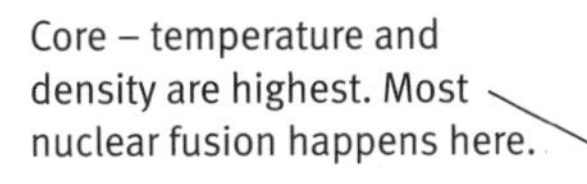

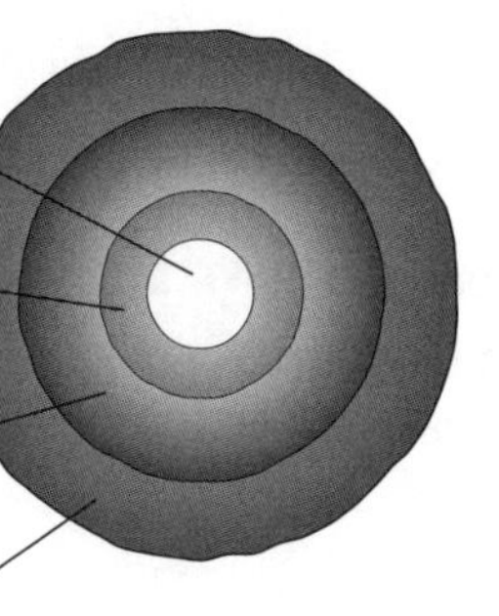

Core – temperature and density are highest. Most nuclear fusion happens here.
Radiative zone – energy is transported outwards from the core by radiation.
Convective zone – convection currents flow here, carrying heat energy to the photosphere.
Photosphere (surface of star) – energy is radiated into space from here.
Energy is transported from the core to the suface by:
• photons of radiation
• convection.

The diagram outlines the processes that happen inside a main-sequence star with a mass that is 5 times the mass of the Sun.

Red giant and supergiant stars

In the core of a main-sequence star, hydrogen nuclei fuse to form helium. Eventually the hydrogen runs out. The pressure decreases. The core collapses.

The outer layers of the star, which contain hydrogen, then fall inwards. New fusion reactions begin in the core. These reactions make the outer layers of the star expand. The photosphere cools, and its colour changes from yellow to red. A **red giant** or **supergiant** has formed.

While the outer layers of a red giant or supergiant expand, its core gets smaller. It becomes hot enough for helium nuclei to fuse together to form heavier nuclei. The more massive the star, the hotter its core and the heavier the nuclei it can produce by fusion.

- In red giants, fusion reactions produce nuclei of carbon. Further fusion reactions then make heavier nuclei such as nitrogen and oxygen.
- In supergiants, the core pressure and temperature are even higher. Fusion reactions may produce elements with nuclei as heavy as that of iron.

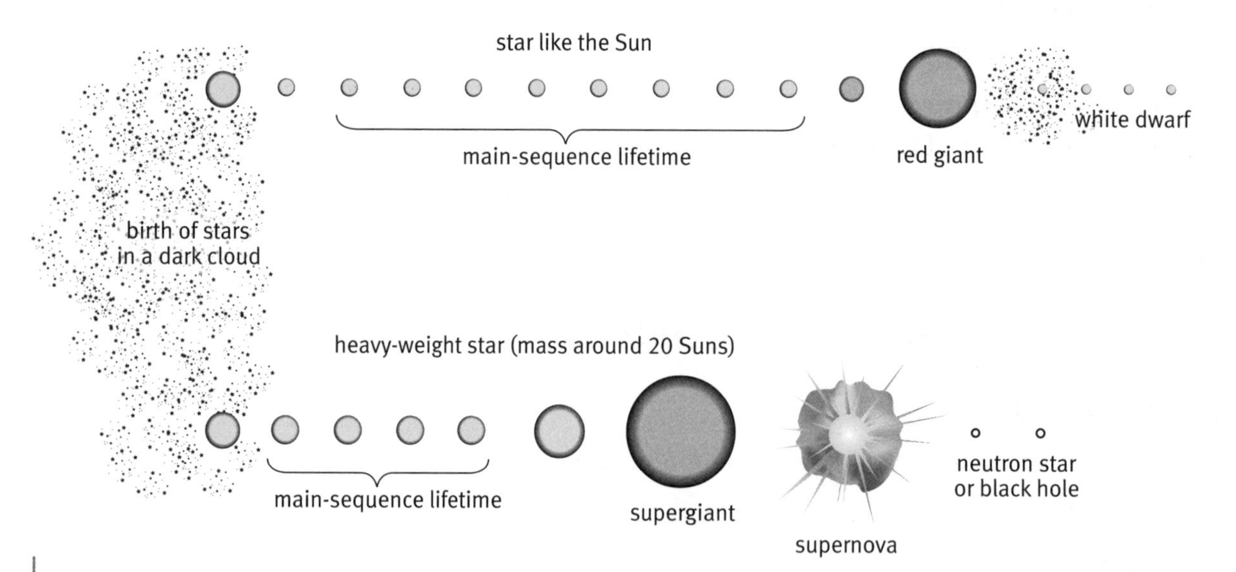

The diagrams show how stars change.

White dwarf stars

The Sun has a relatively low mass. When it becomes a red giant it will not be compressed further once its helium has been used up. The star will shrink to become a **white dwarf star**. There is no fusion in a white dwarf. It gradually cools and fades.

Supernova

When the core of a supergiant is mainly iron, it explodes. This is a **supernova**. It is so hot that fusion reactions produce atoms of elements as heavy as uranium.

After a supernova explosion, a dense core remains.

- A smaller core becomes a **neutron star**.
- A bigger core collapses to become a **black hole**. A black hole has so much mass concentrated into a tiny space that even light cannot escape from it.

The clouds of dust and gas blown outwards by a supernova may eventually form new protostars.

P7.4.22 The Hertzsprung–Russell (H–R) diagram

The H–R diagram is a plot of the luminosity of a star against its temperature. Different types of stars are in different regions of the graph.

For main-sequence stars, the diagram shows a correlation. The hotter the star, the more radiation it emits and so the greater its luminosity.

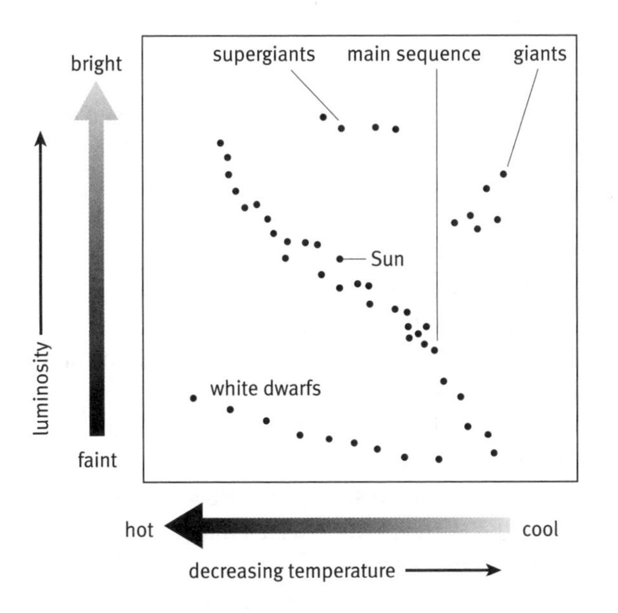

Exam tip

In the exam you may be asked to identify regions on the H–R graph where different types of star are located.

P7.4.29–31 Exoplanets

Astronomers have found convincing evidence of planets orbiting nearby stars. These are **exoplanets.**

Some exoplanets might have conditions that are just right for life. Because of this, some scientists think it is likely that life may exist elsewhere in the Universe.

However, no evidence of extraterrestrial life has so far been detected.

Use extra paper to answer these questions if you need to.

1 The graphs below show the frequency spectra of three stars.

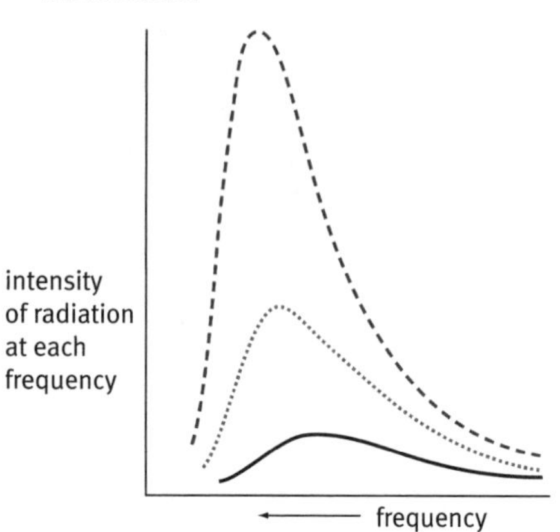

a Label the hottest star and the coolest star.

b The stars are different colours: red, white, and yellow. Label each spectrum with the colour of the star.

2 Complete the table.

Temperature (°C)	Temperature (K)
0	
200	
–200	
	0
	7
	300

3 **a** A scientist heats 100 cm^3 of gas from 200 K to 600 K. Calculate the new volume of the gas.

b A scientist cools 10 dm^3 of gas from 500 K to 250 K. Calculate the new volume of the gas.

4 A scientist fills a container with a gas until its pressure is 1×10^5 Pa. She heats the gas from 300 K to 600 K. Its volume does not change. Calculate the new pressure of the gas.

5 A scientist fills a container with a gas until its pressure is 2×10^5 Pa. She cools the gas from 300 K to 200 K. Calculate the new pressure of the gas.

6 A scientist doubles the volume of a fixed mass of gas, keeping the temperature constant. Write **T** next to the statements below that are true.

a On average, the gas particles get further apart.

b The pressure of the gas will increase.

c Gas particles hit the walls of the container with a greater force.

d Gas particles hit the walls of the container less often.

7 Jasmine writes about the formation of a protostar. She makes mistakes in some sentences. Write a corrected version of each incorrect sentence below.

a In space, a magnetic force compresses a cloud of hydrogen and helium gas.

b The gas particles attract each other.

c The gas particles get closer and closer, and the cloud collapses inwards.

d The volume of the cloud has increased.

e As the gas particles fall towards each other they move more and more slowly, so the temperature and pressure increase.

8 Complete the nuclear equations below.

a $^{12}_{6}C + ^{1}_{1}H \rightarrow ^{...}_{7}N$

b $^{13}_{7}N \rightarrow ^{13}_{...}C + ^{0}_{+1}e$

c $^{13}_{6}..... + ^{...}_{...}H \rightarrow ^{14}_{7}.....$

d $^{...}_{...}N + ^{1}_{1}H \rightarrow ^{12}_{6}C + ^{4}_{2}He$

9 The diagram shows the cross-section of a main-sequence star.

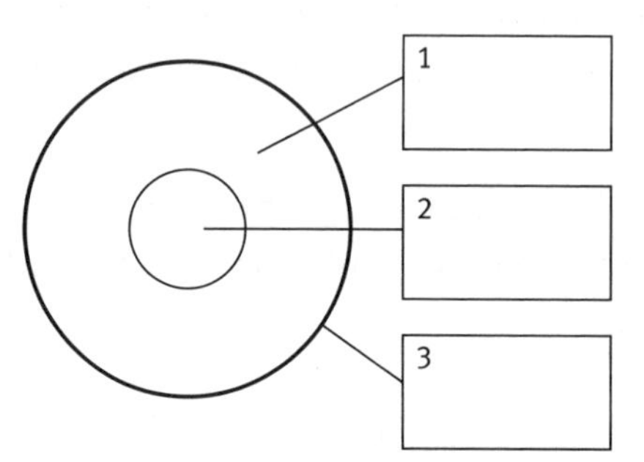

Write the correct letter in each box. You will need to write more than one letter in each box.

a convective and radiative zone

b core

c photosphere

d Most nuclear fusion takes place here.

e Energy radiates into space from here.

f Energy is transported through this part of the star by convection and radiation.

g In this part of the star, the temperature and density are highest.

10 Write the letters of the star types below on the correct places in the Hertzsprung–Russell diagram.

a main-sequence stars

b white dwarfs

c red giants

d supergiants

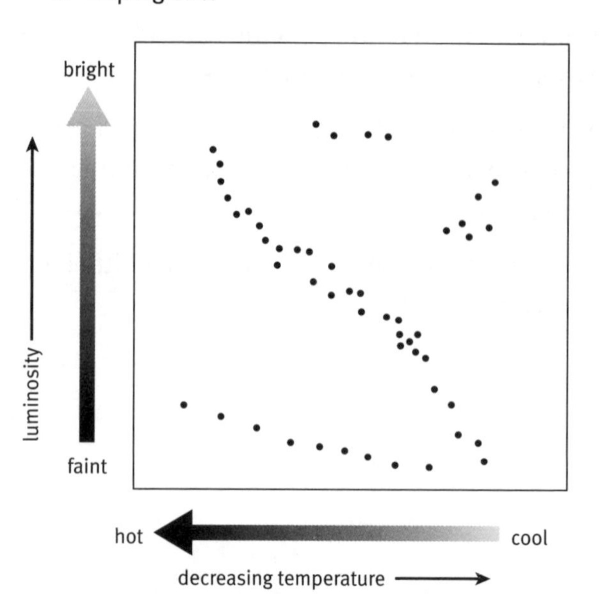

1 X is a star in the Milky Way.

Use the evidence below to suggest what type of star X is most likely to be. Explain your decision.

The quality of written communication will be assessed in your answer to this question.

Write your answer on separate paper or in your exercise book.

Evidence A

Colour of star X	yellow-white
Is star X visible from Earth?	yes

Evidence B

Element	Percentage of element on surface of star X
hydrogen	73.46
helium	24.85
oxygen	0.77
carbon	0.29
iron	0.16
neon	0.12

Evidence C

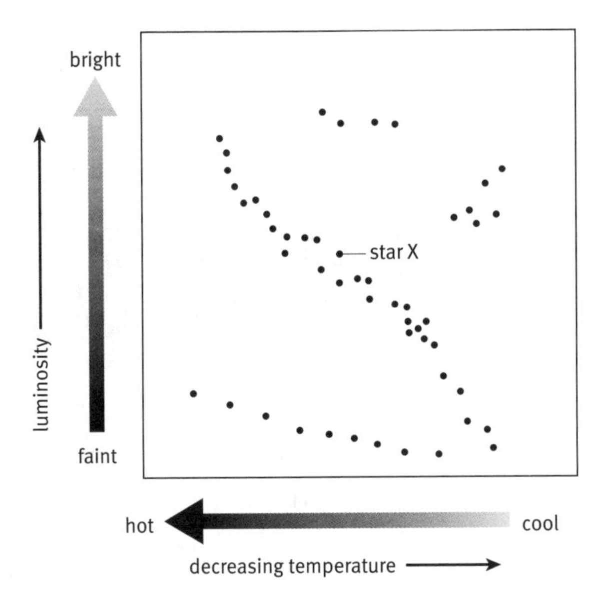

Evidence D

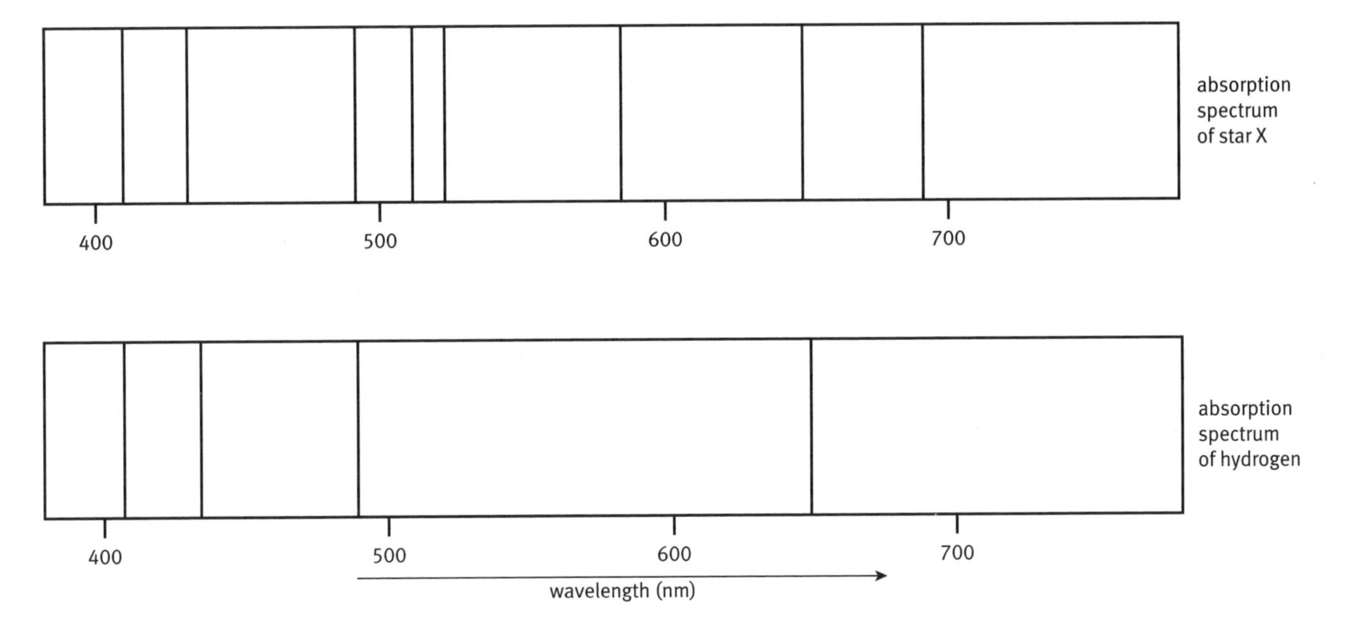

Total [6]

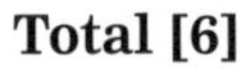

Going for the highest grades

H 2 The equations below show three fusion reactions that take place in the core of the Sun. **The equations are not complete.**

Reaction 1 $^{1}_{1}H + ^{1}_{1}H \rightarrow ^{\cdot}_{1}H + ^{0}_{1}e^{+}$

Reaction 2 $^{1}_{1}H + ^{2}_{1}H \rightarrow ^{3}_{2}He$

Reaction 3 $^{3}_{2}He + ^{3}_{2}He \rightarrow ^{4}_{2}\cdots\cdots + ^{1}_{1}H + ^{1}_{1}H$

a Name the particle represented by the symbol $^{0}_{1}e^{+}$

____________ [1]

b Complete the equation for reaction 1.

Write your answer on the dotted line in the reaction 1 equation above. [1]

c Identify the missing symbol in the reaction 3 equation.

Write your answer on the dotted line in the reaction 3 equation above. [1]

d The table gives the masses of all the particles in reactions 1, 2, and 3.

Formula of particle	Mass of particle (atomic mass units, u)
$^{1}_{1}H$	1.00728
$^{2}_{1}H$	2.01355
$^{3}_{2}He$	3.01493
$^{4}_{2}He$	4.00160
$^{0}_{1}e^{+}$	0

i Calculate the mass changes for reactions 1 and 2.

Give your answers in atomic mass units.

Mass change for reaction 1 = ____________ u

Mass change for reaction 2 = ____________ u [4]

ii The mass change for reaction 3 is 0.01370 u.

Identify the reaction (1, 2, or 3) that releases the most energy.

Explain your choice.

____

____________________________ [2]

e In 1 second, the Sun radiates 3.9×10^{26} J of energy.

Calculate the mass lost by the Sun in 1 second.

The speed of light in a vacuum is 3×10^{8} m/s.

Mass lost by Sun in 1 second = ____________ kg [2]

Total [11]

P7.5 The astronomy community workout

1 Draw lines to join the beginnings of the sentences to the endings.

Draw one, two, or three lines from each beginning.

Beginning
Disadvantages of space telescopes are that they are …
Computers are used to control telescopes because they can …
International cooperation in astronomy allows …
In deciding where to site a new observatory it is necessary to consider …

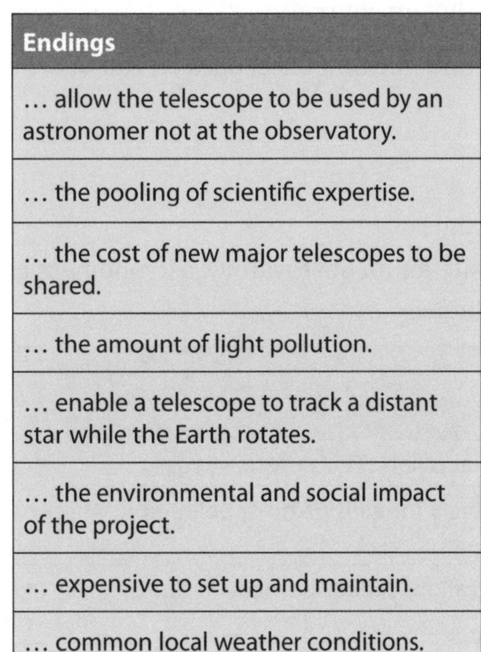

Endings
… allow the telescope to be used by an astronomer not at the observatory.
… the pooling of scientific expertise.
… the cost of new major telescopes to be shared.
… the amount of light pollution.
… enable a telescope to track a distant star while the Earth rotates.
… the environmental and social impact of the project.
… expensive to set up and maintain.
… common local weather conditions.

2 Read the article in the box.

The Royal Greenwich Observatory and Telescopes

In 1675 King Charles II commissioned the building of the Royal Observatory in Greenwich. Its aim was to provide accurate star data for use at sea. Greenwich was a small village outside London, several miles from the smoke of London. The observatory was built in a royal park, so the land did not need to be purchased.

Over the years, many important observations were made from Greenwich. But by 1945 light pollution from the expanding London meant that the site was no longer useful. A new observatory was built in Sussex.

Several telescopes were built in Sussex, including the Isaac Newton Telescope (INT). But the British climate was never ideal, so in 1984 the INT was moved to a high-altitude, clear-sky site in the Canary Islands.

a Underline the phrase in the first paragraph that gives a scientific reason for the location of the observatory.

b Draw a (ring) around the phrase in the first paragraph that gives a reason for the location of the observatory that is not scientific.

c Explain what is meant by light pollution.

____

d Suggest why the British climate might have been a problem for astronomy.

____

e Give two advantages of the INT being on a high-altitude site.

____

P7.5.1–2, P7.5.8 Choosing observatory sites

Astronomers use huge telescopes to collect the weak radiation from faint or very distant sources.

The major optical and infrared telescopes on Earth are in:

- Chile
- Hawaii
- Australia
- the Canary Islands.

When choosing a site for an observatory, astronomers consider the factors in the table.

Exam tip

You may be asked to evaluate telescope sites. If so, compare the advantages and disadvantages of each site. Then state – with reasons – which site you believe to be better.

Astronomical factors	
Factor	**Solution**
The atmosphere refracts light. This distorts images.	A high location (e.g. on a mountain) reduces this problem.
Light is refracted more if the air is damp or polluted.	Locating a telescope in an area with dry, clean air results in higher-quality images.
Astronomical observations cannot be made in cloudy conditions.	Choose an area with frequent cloudless nights.
Cities cause light pollution.	Choose an area far from cities.
Other factors • Cost, including travel to and from the telescope for supplies and workers. • Environmental impact near the observatory. • Impact on local people. • Working conditions for employees.	

P7.5.3–4 Computer control of telescopes

Paolo is an astronomer. He is using a telescope that is thousands of miles away. He uses computer controls to point the telescope towards a particular star and track it as the Earth rotates. Images are recorded digitally and sent electronically to his computer.

Paolo also uses his computer to analyse the images and improve their quality – by adding false colour, for example. Paolo sends his observations to astronomers all over the world.

P7.5.5 Telescopes in space

Telescopes on Earth are affected by:

- the atmosphere, which absorbs most infrared, ultraviolet, X-ray, and gamma radiation
- atmospheric refraction, which distorts images and makes stars 'twinkle'
- light pollution
- bad weather.

All these problems can be overcome by placing the telescope in space, where there is no atmosphere, light pollution, or weather. For example, the **Hubble Space Telescope** has a resolution better than the best Earth-based telescopes.

But there are disadvantages too, including the high costs of setting up, maintaining, and repairing a telescope in space, and uncertainties in future funding.

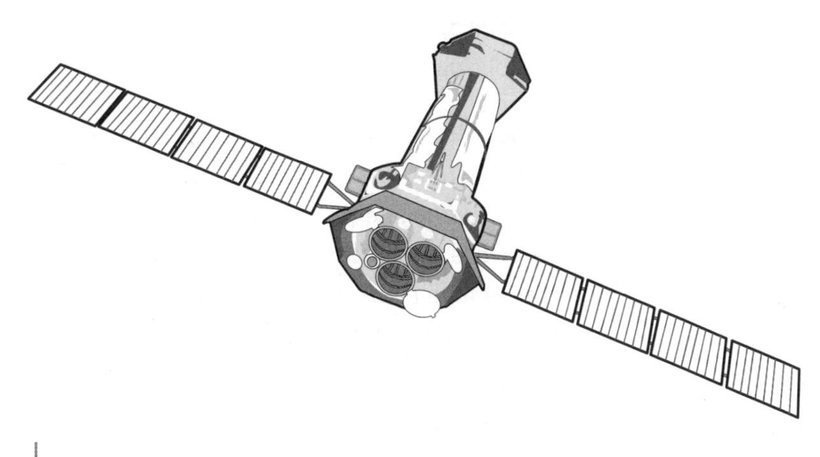

The XMM Newton telescope was launched by the European Space Agency in 1999. It detects X-rays from stars. X-rays from stars cannot be detected on Earth because they are absorbed by our atmosphere.

P7.5.6–7 International collaboration

Collaboration between countries allows the cost of a major telescope to be shared and expertise pooled.

For example, 14 European countries and Brazil have collaborated to run the **European Southern Observatory (ESO)**, which has several telescopes in Chile. Chile provides the base and the office staff. Over 1000 astronomers from all over the world use the facility each year.

The Gran Telescopio Canarias is in the Canary Islands, at the top of a high volcanic peak. It is funded mainly by Spain, with contributions from Mexico and the USA. Planning for the construction of the telescope involved more than 1000 people from 100 companies.

The dome of the Gran Telescopio Canarias.

Exam tip

You need to know two examples showing how international cooperation is essential for progress in astronomy.

Use extra paper to answer these questions if you need to.

1 Write **T** next to the statements that are true. Write corrected versions of the statements that are false.
   a Space telescopes have bigger lenses or mirrors than Earth-based telescopes.
   b The atmosphere refracts visible light.
   c The atmosphere absorbs some types of electromagnetic radiation.
   d Computer control enables a telescope to track a star while the Earth rotates.
   e Light pollution makes it easier to see stars.
   f Space telescopes can detect electromagnetic waves that cannot be detected at ground level.

2 Put ticks in the boxes next to the places below that have major optical and infrared astronomical observatories on Earth.
   a Australia ☐
   b Bangladesh ☐
   c Canary Islands ☐
   d Chile ☐
   e Hawaii ☐
   f UK ☐
   g Zimbabwe ☐

3 Write S next to scientific factors that influence the choice of site for a major astronomical telescope. Write O next to other factors.
   a height above sea level
   b cost
   c travel distance for astronomers
   d number of cloudless nights
   e amount of atmospheric pollution
   f how much water is in the air
   g environmental impact near the observatory
   h working conditions for the people who work at the observatory
   i the availability of staff to build and administer the observatory
   j distance from built up areas that cause light pollution

4 Use the words in the box to fill in the gaps. Each word may be used once, more than once, or not at all.

track **data** **process** **travel** **positioned** **record** **equipment** **communicate**

Using a computer to control a telescope means that astronomers do not have to go to a telescope to collect _______ from it. This saves money on _______. Computers can be used to _______ space objects continuously through the night, and to make sure the telescope is _______ precisely.
Astronomers also use computers to _______ and _______ data, and to _______ their findings with colleagues all over the world.

5 Read the following extract.

> The Hubble Space Telescope (HST) was launched as a joint venture between the European Space Agency in Europe and NASA in the USA. The telescope has a 2.4-m primary mirror, and its instrumentation allows it to work at all frequencies from infrared through to ultraviolet.

   a Is the Hubble Space Telescope a reflector or refractor? Explain how you know.
   b List two advantages of international cooperation in projects such as the Hubble Space Telescope.
   c The Hubble Space Telescope is smaller than many Earth-based telescopes, but it is still able to produce higher-quality images than telescopes on Earth. Explain why.

6 The table gives data about some major reflecting telescopes on Earth.

Name of telescope	Aperture (m)	Paid for by	Site
Gran Telescopio Canarias	10.4	Spain, Mexico, USA	Canary Islands
Keck 1	10	USA	Hawaii
Southern African Large Telescope	9.2	South Africa, USA, UK, Germany, Poland, New Zealand	South Africa
Subaru	8.2	Japan	Hawaii
Lamost	4.9	China	China

   a Which telescope is capable of producing the brightest images of faint or distant sources? Explain your answer.
   b Identify two telescopes that are the results of international cooperation.

7 The table gives the altitude (height above sea level) of four telescopes.

Name of telescope	Height above sea level (m)
Keck 1	4145
Gran Telescopio Canarias	2267
Subaru	4145
Southern African Large Telescope	1783

In which telescope(s) will the effects of the atmosphere on the light being collected be least?

1 Imagine that a group of astronomers in India is deciding where to build an astronomical telescope.

Two possible sites are shown on the map.
The table gives data about the two sites.

	Site 1 – near Leh	Site 2 – near Mumbai
Height above sea level (m)	3500	0
Climate	Low rainfall Very hot in summer Very cold in winter	Hot all year round Dry for 7 months of the year High rainfall June–September
Population of town / city	38 000	20 500 000
Education of local people	75% of adults can read and write	97% of adults can read and write Many computer experts in city
Transport	Main roads to other cities blocked by snow in winter No rail or sea transport	Many major roads to other cities Big railway station and harbour
Other factors	People moving to Leh experience difficulty breathing at first because of its high altitude. Historic city with ancient palace	Many sports and cultural activities in city

a Use the data in the table to evaluate the two sites.

✎ The quality of written communication will be assessed in your answer to this question.

Write your answer on separate paper or in your exercise book.

[6]

b Suggest further data the astronomers could collect and evaluate before making their final decision about the better site for the telescope.

____

____ [2]

Total [8]

2 Astronomers use telescopes outside the Earth's atmosphere to observe distant planets, galaxies, and other astronomical objects.

The tables give data about space telescopes and information about different types of electromagnetic radiation.

Name of telescope	Space agency that launched and maintains telescope	Type of radiation detected by telescope
Hubble	NASA (USA) and European Space Agency	ultraviolet, visible
AGILE	Italian Space Agency	gamma, X-ray
MOST	Canadian Space Agency	visible
Herschel	NASA (USA) and European Space Agency	infrared
RadioAstron	Astron Space Centre (Russia)	radio

Type of radiation	Is this type of radiation absorbed by our atmosphere?	Examples of space objects that emit this type of radiation
gamma	yes	supernovae, neutron stars, black holes
X-ray	yes (over long distances)	black holes, main-sequence stars, remains of supernovae
ultraviolet	yes	Sun, other stars, galaxies
visible	no	stars, galaxies
infrared	yes	cooler stars, redshifted galaxies
radio	no	remains of supernovae

a List four types of object that can be observed by the AGILE space telescope.

____

____ [4]

b Name two telescopes that an astronomer might use to observe the remains of supernovae.

____ [2]

c Name two telescopes that are the result of international collaboration.

____ [2]

d Identify two advantages of international collaboration in astronomical research.

____

____ [2]

Total [10]

Data: their importance and limitations

7
Overall, the scientists claim that their results are **repeatable**. The same scientists, using the same equipment and sending the neutrinos on the same journey, got similar results many, many times.
The scientists spent months doing checks, and didn't find any errors.
8
The scientists must have calculated the **mean** time for all those journeys. The mean gives the best estimate of its true value.
9
What about the **range**? Isn't that important?
10
Yes, the range is the maximum and minimum value of a variable.
For the set of repeated time measurements, the true value probably lies within the range of the journey times measured.
11
H
Now the scientists want to know if their results are **reproducible**. Different scientists based in Japan and the USA will do similar investigations. Will their neutrinos travel faster than light?
12
H
If different scientists using different equipment get similar results, the measurements are reproducible.
And we can be more confident that the explanation is correct... Neutrinos travel faster than light.

Data: their importance and limitations

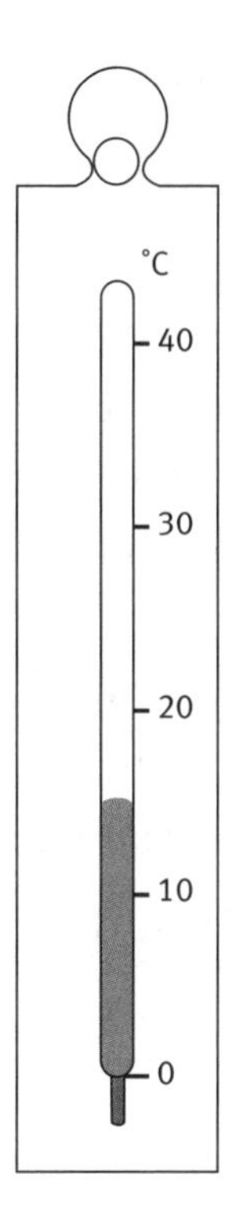

1 The ideal air temperature of a baby's bedroom is 18 °C. A father is worried that the room is too cold, so he hangs a mercury thermometer on the wall between the curtain and the window. He uses the thermometer to measure the temperature, and gets a reading of 15 °C.

Give three reasons why the measurement may not give you the correct value for the air temperature in the room.

____

____

____

____

____

____

2 Draw lines to match each word or phrase to its definition.

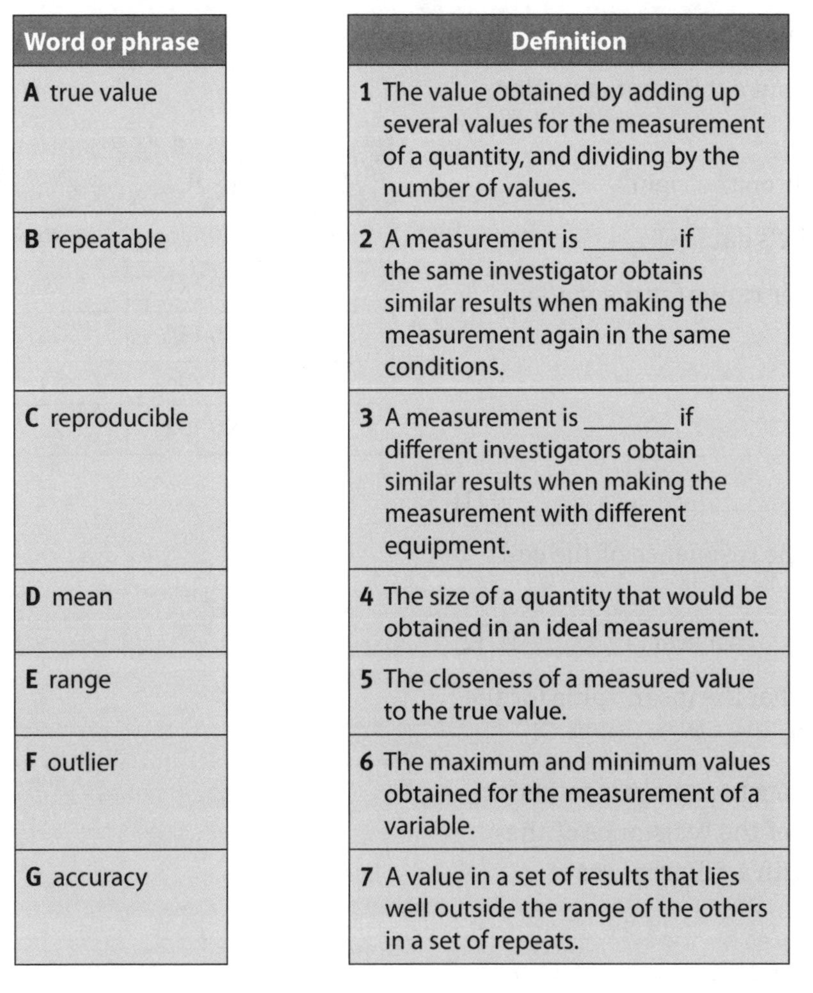

Word or phrase	Definition
A true value	**1** The value obtained by adding up several values for the measurement of a quantity, and dividing by the number of values.
B repeatable	**2** A measurement is ________ if the same investigator obtains similar results when making the measurement again in the same conditions.
C reproducible	**3** A measurement is ________ if different investigators obtain similar results when making the measurement with different equipment.
D mean	**4** The size of a quantity that would be obtained in an ideal measurement.
E range	**5** The closeness of a measured value to the true value.
F outlier	**6** The maximum and minimum values obtained for the measurement of a variable.
G accuracy	**7** A value in a set of results that lies well outside the range of the others in a set of repeats.

Data: their importance and limitations

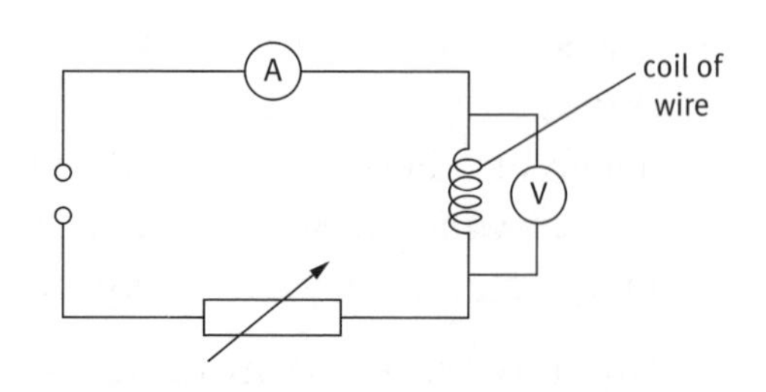

1 Tillie wants to find out the resistance of a coil of wire.

She sets up the circuit opposite.

Tillie records values for voltage and current.

She adjusts the variable resistor, and records the voltage and current again.

She repeats until she has collected the data in the table on the right.

Voltage (V)	Current (A)	Resistance (Ω)
1.5	0.060	
3.0	0.115	26.0
4.5	0.188	24.0
6.0	0.240	25.0
7.5	0.234	

a Use the equation $R = V/I$ to calculate the missing values for resistance.

Write your answers in the table. [1]

b **i** Identify the outlier from all the values for R in the table.

______________________________ [1]

ii Suggest what Tillie should do about the outlier you have identified.

______________________________ [1]

c Charlie does the same investigation as Tillie.

She uses the same coil of wire, but a different ammeter, voltmeter, and variable resistor.

Charlie's results are in the table on the right.

Voltage (V)	Current (A)	Resistance (Ω)
3.0	0.115	26
3.5	0.121	29
4.0	0.154	26
4.5	0.196	23
5.0	0.192	26

i Compare Tillie's and Charlie's data sets.

Which data set has the bigger range for the voltage values?

Explain how you decided.

____

______________________________ [1]

ii Estimate the true value of the resistance of the coil from Charlie's data.

Answer = ________ Ω [2]

H **d** Tillie calculates from her data that the mean value for the resistance of the coil was 25.0 Ω.

Use Tillie's and Charlie's data sets to explain whether or not you think the true value of the resistance of the coil changed when measured with a different set of equipment.

____

____

____

______________________________ [3]

Total [9]

Cause–effect explanations

1
Miss Vine told us to investigate factors that affect the focal length of lenses.
I know that focal length is the distance from a lens to the sharpest image of a faraway object. But what shall we do?
2
Well, the **outcome** is the focal length. We can easily measure that.
What **factors** will affect the outcome?
3
I think the fatness of a lens will affect the outcome. And maybe the type of glass. Those are the factors.
4
Let's investigate the effect of changing lens fatness.
To make it a **fair test** we need to control the other factor – the type of glass. We'll need to make sure this is the same for all the lenses we use. If we don't the design of our investigation will be flawed.
5
One hour later...
OK. The data show a **correlation**. The fatter the lens, the shorter its focal length. Brilliant!
6
But maybe the change in focal length is caused by some other factor. A correlation does not always mean there is a **causal link**.
True. But I can't think of another factor that would change both lens fatness and focal length. Can you?

Cause–effect explanations

1 Use the clues to fill in the grid.

1. A correlation between a factor and an outcome does not necessarily mean that there is a c... link.
2. Scientists often think about processes in terms of factors that may affect an o...
3. An input variable is also called a f...
4. In an investigation, if you control all factors that may affect an outcome except the factor you are investigating, you are doing a f... test.
5. When there is evidence that a factor is correlated with an outcome, scientists will only accept that the factor causes the outcome if they can think of a sensible m... to link the factor and outcome.
6. If you are investigating the effect of a factor on an outcome, and you do not control other factors, your investigation design is f...
7. Smoking increases your c... of getting lung cancer.
8. To investigate the claim that a factor increases the chance of an outcome, scientists may compare samples that are m... on as many other factors as possible.
9. If an outcome variable i... as a factor increases, there is a correlation between the factor and the outcome variable.
10. When investigating whether mobile phone use increased the chance of cancer, scientists compared r... samples so that other factors were equally likely in both samples.
11. If you are investigating the effect of the number of coils on generator voltage, you must c... other factors such as speed of turning and whether or not there is an iron core.

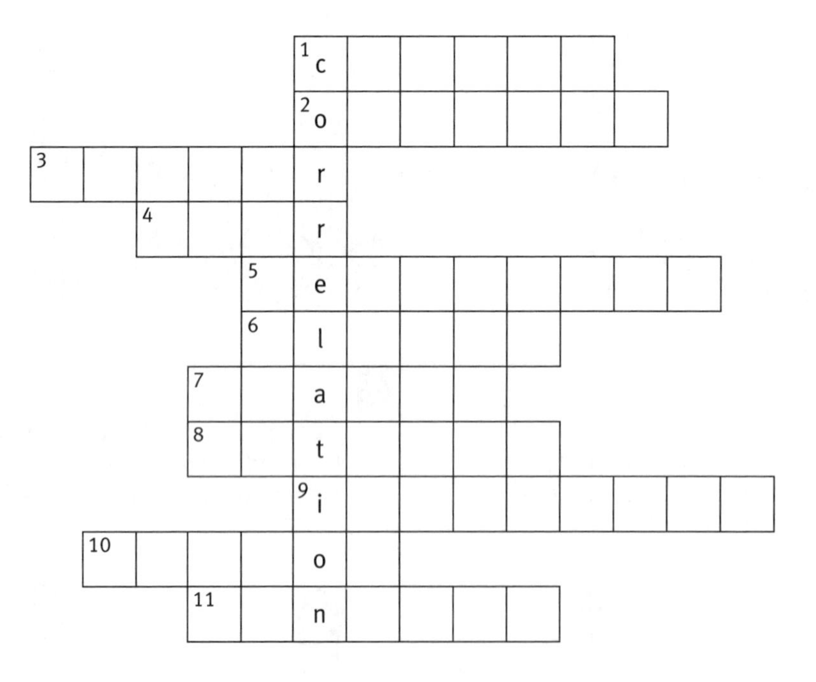

Cause–effect explanations

1 Sam is investigating gases. He wants to find out how changing the temperature of a gas affects its volume.

a Identify the outcome in the investigation.

____ [1]

b **i** Identify three factors that may affect the outcome.

____ [1]

ii Which of these factors should Sam control?

____ [1]

iii Explain why Sam should control these outcomes.

____ [1]

c Sam sets up the apparatus opposite.

He records the height of the column of air in the tube at five different temperatures.

He collects the data in the table.

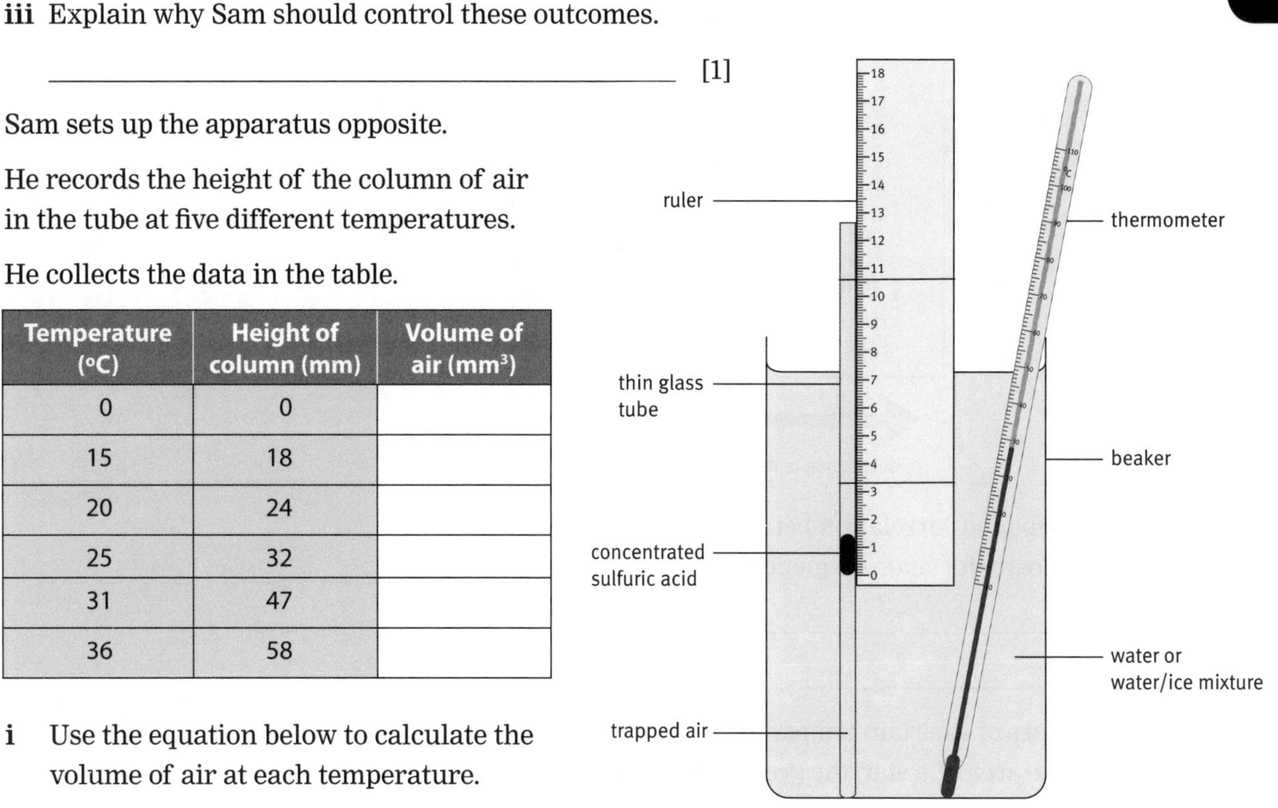

Temperature (°C)	Height of column (mm)	Volume of air (mm^3)
0	0	
15	18	
20	24	
25	32	
31	47	
36	58	

i Use the equation below to calculate the volume of air at each temperature.

Write your answers in the table above.

volume = $80\ mm^3$ + [height (mm) × cross-sectional area of tube (mm^2)]

volume = $80\ mm^3$ + [height (mm) × 0.8 (mm^2)] [2]

ii Use the data in the table to plot a graph to show the relationship between temperature and gas volume.

Use a separate piece of graph paper. [3]

iii Describe the relationship shown by the graph.

____ [1]

iv Use ideas about particles to explain the relationship shown on the graph.

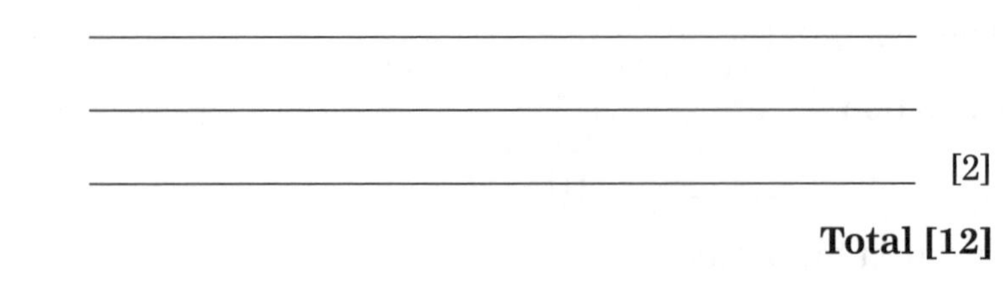

[2]

Total [12]

2 Use the Hertzsprung–Russell diagram below to help you answer this question.

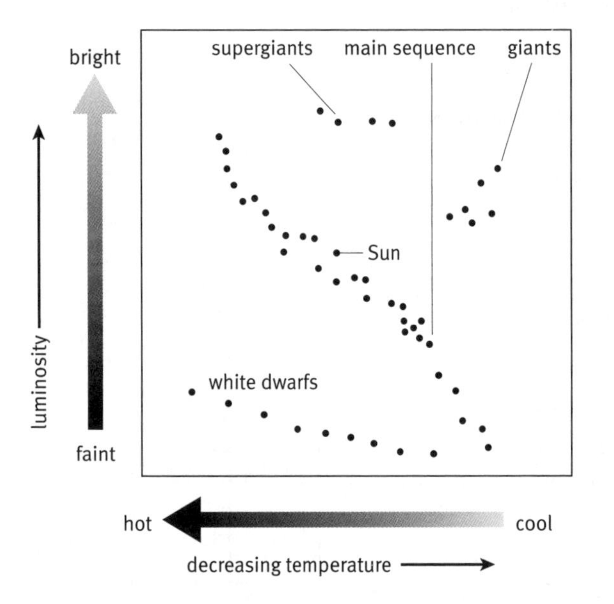

a Describe the correlation between temperature and luminosity for main-sequence stars.

____

____ [1]

b For stars of a certain temperature, the greater the surface area of a star, the greater its luminosity.

i Identify the outcome and input factor in the statement above.

____

____ [2]

ii Use the statement at the start of part (b) to explain the relative positions of red giants and main-sequence stars on the Hertzsprung–Russell diagram.

____

____

____ [2]

Total [5]

Developing scientific explanations: mapping the seafloor

1 It's 2010. A team of geologists are on a boat, north of Scotland. They send sound waves more than 2 km below the seafloor, and analyse the echoes that come back up to the boat.

We're certainly not looking at boring layers of mud and sand. Let's map the data. Then maybe we can come up with an explanation.

This data is amazing. I've never seen anything like it. In some places, the bottom of this mudstone layer is 1 km below the seafloor. But in other places, the mudstone goes down much deeper.

2 One year later, at a scientific conference.

I mapped the data. The shape of the rock under the mudstone, 2 km below the seafloor, looks like the surface of land, with hills, valleys, and rivers. We ***thought creatively*** *about the data, and came up with an* ***hypothesis – an explanation that might account for the data.***

3

Here's our hypothesis: 56 million years ago, seafloor rock was forced upwards, above the surface of the sea. For two million years, wind and water shaped the landscape. Then the land sank down again, beneath the sea. Here, it was covered with sediment, which formed new layers of rock that buried the landscape.

We think the landscape rock was forced up and then down by hot mantle material moving beneath the seafloor tectonic plate.

4

From our hypothesis, we made a ***prediction.*** *There are fossils of land plants in the rock that was once land. There are fossils of sea animals in the layers of rock above and below the rock that was once land.*

5

An oil company drilled through the rock, and gave us samples. Our prediction was correct. This made us ***more confident in our explanation.*** *Of course, we* ***cannot prove*** *our explanation is correct.*

Developing scientific explanations: fossil skull

1 It's 1924. Two scientists are reporting their latest fossil find.

This is it. A fossilised skull from the first human species: australopithecine.

Before we found the skull, we were confident in the explanation that human ancestors developed big brains before they walked upright. We predicted that any fossils we found would support this explanation.

2

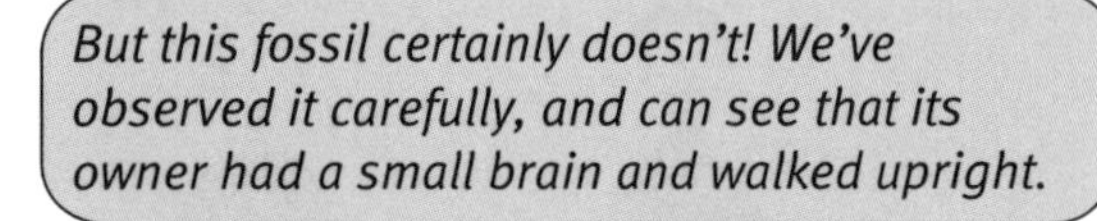

But this fossil certainly doesn't! We've observed it carefully, and can see that its owner had a small brain and walked upright.

*So our **observations of this new find disagree with our prediction**. We're now much **less confident in the explanation** that humans had big brains before they walked upright.*

Developing scientific explanations

1 Write **T** next to the statements that are true and **F** next to the statements that are false.

a Scientific explanations simply summarise data. ___

b If the data agrees with an explanation, the explanation must be correct. ___

c Developing a scientific explanation requires creative thought. ___

d A scientific explanation must account for most, or all, of the data already known. ___

e An explanation should always explain a range of phenomena that scientists didn't know were linked. ___

f Scientists test explanations by comparing predictions based on them with data from observations or experiments. ___

g If an observation agrees with a prediction that is based on an explanation, it proves that the explanation is correct. ___

h If an observation does not agree with a prediction that is based on an explanation, it decreases confidence in the explanation. ___

i If an observation does not agree with a prediction that is based on an explanation, then the observation must be wrong. ___

j If an observation does not agree with a prediction that is based on an explanation, then the explanation must be wrong. ___

Now write corrected versions of the **six** false statements on the lines below.

____

____

____

____

____

____

____

____

Developing scientific explanations

1 Scientists collected these data:

- All around the world, rocks from 65 million years ago contain high levels of iridium.
- Asteroids contain iridium.

From the data, they developed this **explanation**.

An asteroid hitting the Earth 65 million years ago caused rocks of that age to contain high levels of iridium.

Other scientists used the explanation and extra data to make a prediction:

Extra data: Asteroids and comets were made at the same time.

Prediction: Comet tail dust contains iridium.

The scientists use a spacecraft to collect comet tail dust.

They will examine the dust to find out if it contains iridium.

Tick the boxes next to the **two** statements that are true.

If they find iridium, this will increase confidence in the explanation. ☐

If they find iridium, this will prove the explanation is correct. ☐

If they do not find iridium, this will show that the prediction is definitely wrong. ☐

If they do not find iridium, the prediction may still be correct. ☐

Total [2]

2 Nearly 2000 people who lived near Lake Nyos in Cameroon died in 1986.

a Scientists wanted to find out why the people died. They collected these data:

A Carbon dioxide is soluble in water.

B If carbon dioxide takes the place of air, people die from lack of oxygen.

C There is a volcano below Lake Nyos.

D Carbon dioxide gas is denser than air.

E Sometimes there are small Earth movements near Lake Nyos.

F If you shake a saturated solution of a gas, some of the gas escapes from solution.

G Carbon dioxide has no smell.

H Magma contains dissolved carbon dioxide.

I Carbon dioxide gas is invisible.

i The scientists used their data to develop an explanation. The explanation is in six parts.

Next to each part of the explanation, write one or two letters to show which data each part of the explanation accounts for.

One has been done for you.

Do not write in the shaded boxes.

Part of explanation	Data that this part of the explanation accounts for	
1 Carbon dioxide gas bubbles into the bottom of Lake Nyos.	C	
2 Carbon dioxide dissolves to make a saturated solution.		
3 There was a small Earth movement. This released 80 million cubic metres of carbon dioxide gas from the lake.		
4 Carbon dioxide gas filled the valleys around the lake.		
5 No-one detected the carbon dioxide gas, so no-one ran away.		
6 1700 people died.		

[4]

ii The explanation accounts for all the data. Does this mean that the explanation must be correct? Give a reason for your decision.

____

____ [1]

b Scientists predict that Lake Nyos will release carbon dioxide again in future. They expect more people to die. They do not know when this will happen.

Suggest one reason why the scientists cannot know when Lake Nyos will next release a large amount of carbon dioxide.

____

____ [1]

Total [6]

3 In the 1930s, scientists investigated the effects of bombarding uranium with neutrons.

a Some of the following statements report **data** collected by the scientists, and one is a possible **explanation**.

Write a **D** next to each of the three statements that are data.
Write an **E** next to the one statement that is an explanation.

Statement	D or E
1 On bombarding uranium with neutrons, Enrico Fermi observed that at least four radioactive substances were produced.	
2 Hahn and Strassmann bombarded uranium with neutrons. At least three substances were formed that had chemical properties similar to barium.	
3 Hahn and Strassmann could not separate the substances with properties similar to barium from barium itself.	
4 Isotopes of barium are formed when uranium is bombarded with neutrons.	

[2]

b Scientists Lise Meitner and Otto Frisch read about the work described in the table above. They came up with a new explanation:

The uranium nucleus is unstable. When bombarded with neutrons, it divides itself into two nuclei of roughly equal size.

Suggest why Lise and Otto must have used creative thinking to help them develop this explanation.

____

____ [1]

c Lise and Otto used their explanation to make a prediction:

The nuclei formed when a uranium nucleus breaks down will be unstable. These nuclei will decay to form lighter atoms.

Later observations showed that the prediction was correct.

Tick the boxes next to the **two** statements that are true.

- The fact that the prediction is correct proves the explanation is correct. ☐
- The fact that the prediction is correct increases confidence in the explanation. ☐
- If the prediction was wrong, we would be less confident in the explanation. ☐
- If the prediction was wrong, we would be sure the explanation was wrong. ☐

[2]

d Lise and Otto explained that when a uranium nucleus decays, some of its mass is converted into energy.

Describe an observation that would support this explanation.

____

____ [1]

Total [6]

4 In the 1630s, scientist Galileo investigated acceleration.

He set up the apparatus below.

He measured the times for the ball to roll down different sections of the slope.

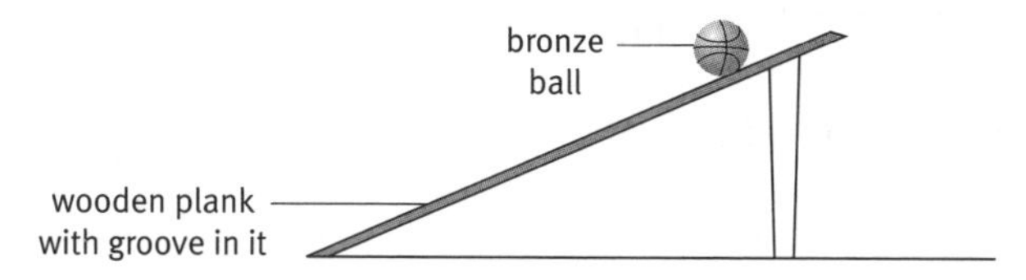

a The **statements** describe what Galileo did.

The **flow chart** shows how science explanations develop.

Write the letter of each statement in an empty box on the flow chart. Some boxes need more than one letter.

Statements

A He measured the time for the ball to roll down the whole length of the slope.

B He measured the time for the ball to roll down half the length of the slope.

C He measured the time for the ball to roll down a quarter the length of the slope.

D He calculated the acceleration of the ball for each of its journeys.

E The acceleration of the ball is the same, no matter what distance it travels.

F He thought that if he rolled the ball down a steeper slope, the acceleration of the ball would be the same, no matter what distance it travelled.

G He measured the times for the ball to roll down different lengths of a steeper slope.

Flow chart

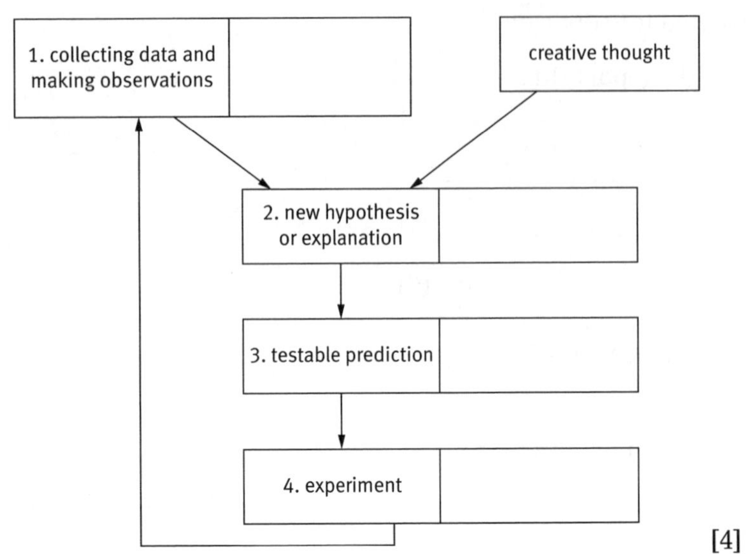

[4]

b Galileo made the groove in the apparatus as smooth as possible. Suggest why.

____

____

____________________________________ [1]

c Galileo repeated his measurements many, many times. Suggest why.

____

____

____________________________________ [1]

d Galileo came up with another hypothesis:

The steeper the slope, the greater the acceleration of the ball.

i Describe how he might have tested this hypothesis.

____

____

____

____________________________________ [2]

ii If the hypothesis was correct, what would his results have shown?

____

____

____

____________________________________ [2]

Total [10]

The scientific community

THE HISTORY PROGRAMME 23/08/2012
Presenters' script

Presenter 1 (Simon)
Welcome to *The History Programme*. This year, 2012, is the hundredth anniversary of Wegener's theory of continental drift. We all now recognise the importance of his ideas. But a century ago scientists laughed at him.

Presenter 2 (Janet)
Yes, that's right. Wegener explained that the east coast of South America was once joined to Africa's west coast. The two continents had been slowly moving apart ever since. Wegener had lots of data to support his explanation: the shapes, rock types, fossils, and mountain ranges of the two continents matched up closely.

Presenter 1
So why did other scientists disagree with Wegener?

Presenter 2
Well, of course you can't always deduce explanations from data. So it's quite reasonable for different scientists to come to different conclusions, even if they agree about some of the evidence. But there's more to it than that.

Presenter 1
Tell me more.

Presenter 2
It seems that other scientists simply couldn't imagine how massive continents could move across the planet. It was an **idea outside their experience**. Also, they didn't much respect Wegener – he was never regarded as a **member of the community of geologists**.

Presenter 1
And I suppose scientists don't give up their 'tried-and-tested' explanations easily?

Presenter 2
Exactly. Scientists often feel that it's safer to stick with **ideas that have served them well in the past**. Of course, new data that conflicts with an explanation makes scientists stop and think – but it could be that the data is incorrect, not the explanation! Generally, scientists only abandon an established explanation when there are really good reasons to do so, like someone suggesting a better one.

The scientific community

1 The statements below describe some of the steps by which scientists may accept a new scientific claim or explanation.

Write each statement in a sensible place on the flow chart below.

A Many other scientists read the paper. Some may try to reproduce its findings.

B A small number of other scientists read the paper to check the methods and claims, and to spot any mistakes.

C He makes a claim or creates a new explanation.

D The scientist makes corrections to his paper.

E He repeats the investigation to check that he can replicate his own findings.

F Other scientists are sceptical about the new claim or explanation.

G The paper is published in a scientific journal, in print and online.

H Other scientists are more likely to accept the new claim or explanation.

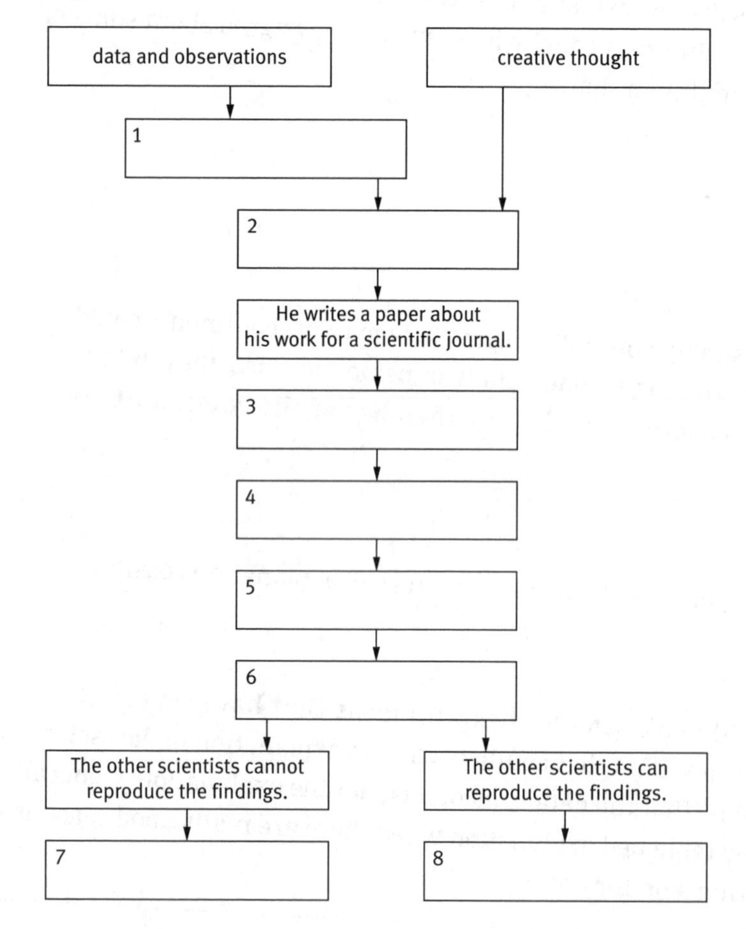

2 Draw a line to link two words or phrases on the circle. Write a sentence on the line saying how the two words are connected.

Repeat for as many pairs as you can.

new scientific claim

peer review

replicate

report

scientific community

unexpected findings

reproduce

scientific journal

critical evaluation

established scientific claim

scientific conference

sceptical

The scientific community

1 a Tick the **two** reasons below that best explain why a scientist may decide to tell others about their research findings at a scientific conference.

So that other scientists can talk about the findings on television. ☐

So that other scientists can write about the findings in a scientific journal. ☐

So that other scientists can try to reproduce the findings in future. ☐

So that other scientists can ask questions about the findings. ☐ [2]

b In 1912, Alfred Wegener presented his idea of continental drift to other scientists at a conference in Germany.

i List three pieces of evidence Wegener used to support his idea.

____

____

______________________________________ [3]

ii Explain why most scientists at the conference did not accept Wegener's idea.

____

____

______________________________________ [3]

iii Today, most geologists accept the scientific explanation that continents drift because of the movement of tectonic plates.

Tick the **two** reasons that best explain why geologists now accept this explanation, even though it was rejected in 1912.

Data collected since 1912 supports the explanation. ☐

The Earth's crust is made up of about 12 tectonic plates. ☐

The Earth's crust is thin compared to the mantle and the core. ☐

Data collected since 1912 agrees with predictions based on the explanation. ☐

[2]

Total [10]

Exam tip

If a question is worth three marks, try to make at least three points in your answer.

Risk: sunbathing and skin cancer

Why are you lying out there in the sun at midday? Where's your sunscreen? It's dangerous to sunbathe – ultraviolet radiation from the Sun hugely increases your chance of getting skin cancer. Each year, nearly 6000 British people get melanoma skin cancer. And skin cancer kills.

Yes, but there are ***benefits*** *too. Sunlight helps you make vitamin D. You need this to strengthen your bones and muscles, and to boost your immune system. Anyway, I feel more confident when I've got a tan. And it's lovely and warm out here...*

OK. I guess it's up to you. No-one is forcing you to take the risk. But think of the consequences of getting skin cancer. And the costs to the NHS of treating you. We'll all pay in the end.

Risk: nuclear power

Risk: thallium scan

7
Can I be sure that the risks from this scan are as low as possible?
We use only a little radiation. It's like having a few X-rays, or a few years of normal background radiation.
8
I guess there are rules about radiation levels for these scans.
You're right. We must use as little radiation as possible to do the job.
9
Would I be contaminated forever?
Thallium-201 has a half-life of 73 hours. Its activity is one quarter of its original value in just over a week. Thallium might come out when you go to the loo, too.
10
Isn't this another risk? What can I do about it?
You need to flush twice, and wash your hands properly. That will reduce the risk to you and others.
11
I don't know what to do. Are you saying I must have the scan?
Of course not. It's up to you.
12
It's a hard decision. The consequences could be long-lasting, if I get cancer. But it is my choice... OK. I'll have the scan.
Good decision. I really believe that the benefits outweigh the risks.

Risk: irradiate spice

Risk

1 Use the clues to fill in the grid.

1 Everything we do carries a risk of accident or h . . .

2 We can assess the size of a risk by measuring the c . . . of it happening in a large sample over a certain time.

3 New vaccines are an example of a scientific a . . . that brings with it new risks.

4 Radioactive materials emit i . . . radiation all the time.

5 The chance of a nuclear power station exploding is small. The c . . . of this happening would be devastating.

6 Governments or public bodies may have to a . . . what level of risk is acceptable in a particular situation.

7 Some decisions about risk may be c . . ., especially if those most at risk are not those who benefit.

8 Sometimes people think the size of a risk is bigger than it really is. Their perception of the size of the risk is greater than the s . . . calculated risk.

9 Nuclear power stations emit less carbon dioxide than coal-fired power stations. Some people think that this b . . . is worth the risk of building nuclear power stations.

10 Many people think that the size of the risk of flying in an aeroplane is greater than it really is. They p . . . that flying is risky because they don't fly very often.

11 It is impossible to reduce risk to zero. So people must decide what level of risk is a . . .

2 Draw a line to link two words on the circle.
Write a sentence on the line saying how the two words are connected.
Repeat for as many pairs as you can.

risk

safe

benefits

chance

consequences

balance

unfamiliar

scientific advances

long-lasting effects

statistically estimated risk

perceived risk

controversial

Risk

1 A group of friends are talking about using mobile phones.

Catherine: I use my phone all the time to talk to family and friends. I'm not worried about microwave radiation from the phone. My skull absorbs most of it.

Zion: I was worried about microwaves from my phone heating up my brain cells and damaging them. So now I use my hands-free set.

Sarah: Running for half an hour heats your brain up more than mobile phone radiation. I won't use my phone any less.

Junaid: I read a 2011 report from the World Health Organization. It says that mobile phone radiation may cause cancer. So I text everyone now.

a Which two people are taking action to reduce the risks from exposure to radiation from mobile phones?

Put ticks (✔) in the boxes next to the two correct names.

Catherine ☐

Sarah ☐

Zion ☐

Junaid ☐ [1]

b Which three people are talking about the hazards from exposure to radiation from mobile phones?

Put ticks (✔) in the boxes next to the three correct names.

Catherine ☐

Sarah ☐

Zion ☐

Junaid ☐ [1]

c Which person mentions something that prevents mobile phone radiation reaching the brain?

Put a tick (✔) in the box next to the correct name.

Catherine ☐

Sarah ☐

Zion ☐

Junaid ☐ [1]

Total [3]

2 Pete lives near a nuclear power station.
He is offered a job in the office at the power station.
He needs to decide whether or not to accept the job.

a The annual risk of a nuclear power station worker developing cancer from exposure to radiation is 0.1%. This is 40 times greater than for a member of the public.

Different people have different opinions about Pete's increased cancer risk.

Draw lines to match each opinion to the person most likely to have the opinion. [2]

Person
Government energy minister
Pete
Jake (Pete's son)

Opinion
The calculated risk of his getting cancer is small, but if it happened the consequences would be terrible.
Nuclear power stations provide electricity for many people. This is worth the small extra cancer risk to power station workers.
The extra cancer risk is significant, but there are few jobs in the area and the pay is good.

b In nuclear power stations, precautions are taken to reduce the risks to workers. Identify one of these precautions.

______________________________ [1]

c If there is a leak of radioactive iodine from a UK nuclear power station, children and pregnant women living nearby may be given potassium iodide tablets. The tablets help to prevent thyroid cancer.

Suggest why the tablets are given only to those living closest to the power station, not to everyone in the UK.

______________________________ [1]

d Kara is Pete's wife. She thinks that UK nuclear power stations should be closed down. She says:

> They should build more nuclear power stations in France. Then the UK could get all its electricity from these power stations. We already import some electricity from France. There is a cable under the sea linking the two countries.

Discuss the risks and benefits of Kara's idea, and identify how different groups of people would be affected by it.

____

____

____

______________________________ [3]

Total [7]

5

3 Barbara and Tom live in the USA. They are discussing risk.

Tom The risk of nuclear power station workers getting cancer as a result of radiation exposure is less than the risk of aeroplane pilots getting cancer as a result of radiation exposure.

Barbara That's not right. I go on aeroplanes all the time. But I would never visit a nuclear power station – think of all that invisible radiation! Nuclear workers must be at greater risk from cancer.

Job	Typical radiation dose (mSv/year)
nuclear power station worker	2.4
pilot who does not fly over the Arctic	5
pilot who regularly flies over the Arctic	9

a Who is correct, Tom or Barbara?

Use data from the table to support your answer.

____

____ [2]

H **b** Suggest two reasons for Barbara's opinion.

____

____ [2]

Total [4]

4 Read the information in the box, then answer the questions that follow

The Hubble Space Telescope

The Hubble Space Telescope (HST) was launched in 1990. Scientists have used observations from the HST to find out more about Cepheid variable stars, black holes, and the expansion of the Universe. The HST has sent amazing images to Earth, including some showing the collision of a comet with Jupiter.

There have been five servicing missions to the HST to repair damaged instruments and install new ones. On these missions, astronauts travelled to the HST by space shuttle. Once there, they went on spacewalks to service the telescope.

The final servicing mission, of 2009, was delayed by several years. The delay followed the disaster in which Space Shuttle Columbia broke up during re-entry to the Earth's atmosphere, killing all seven crew members.

No more servicing missions to the HST are planned. The telescope is expected to continue operating until 2013.

a Identify two benefits of the Hubble Space Telescope.

____

____ [2]

b **i** Identify one risk to astronauts of servicing the Hubble Space Telescope.

____ [1]

ii Suggest one group of people who are likely to think that this risk is worth taking.

____ [1]

iii Suggest one group of people who might think that this risk is not worth taking.

____ [1]

c Before the final servicing mission, the leaders of the Space Shuttle programme considered different options for servicing the HST. Three of these are listed below.

Option 1: Send astronauts to the HST by Space Shuttle. Make sure the shuttle can reach the International Space Station if an in-flight problem develops that would prevent a safe return to Earth.

Option 2: Develop the technology needed to send a robot to service the HST. This would take several years.

Option 3: Do not service the HST again. Allow it to operate for as long as possible without servicing.

Evaluate the risks and benefits of the three options.

✎ The quality of written communication will be assessed in your answer to this question.

Write your answer on separate paper or in your exercise book. [6]

Total [11]

5 Read the article in the box, then answer the questions that follow.

> The Implantable Miniature Telescope is a new device designed to improve sight for people with a disease called AMD. People with AMD cannot recognise faces or watch television because the centre of the retina at the back of their eye is damaged.
>
> The miniature telescope is inserted into the eye in an operation. It has two tiny lenses at each end of a small tube. The telescope magnifies images and projects them onto a healthy part of the retina.
>
> During trials of the device, the vision of 90% of patients improved. However, various risks were identified. Some patients found it very difficult to get used to the device. Other patients suffered damage to the cornea (the transparent covering at the front of the eye). Data from the trial showed that the risk of suffering some damage to the cornea during the first 5 years after having the implant is 20%. The risk of needing a cornea transplant during this time is 4%.

a Suggest one reason that a person may give for having the implant, even though there are risks associated with it.

______________________________ [1]

b Ethel's optician suggests she has an implant.
Ethel says, "I am not going to have the implant.
The risks are too great."

Suggest two reasons for Ethel's decision.

____

____

______________________________ [2]

Total [3]

Making decisions about science and technology

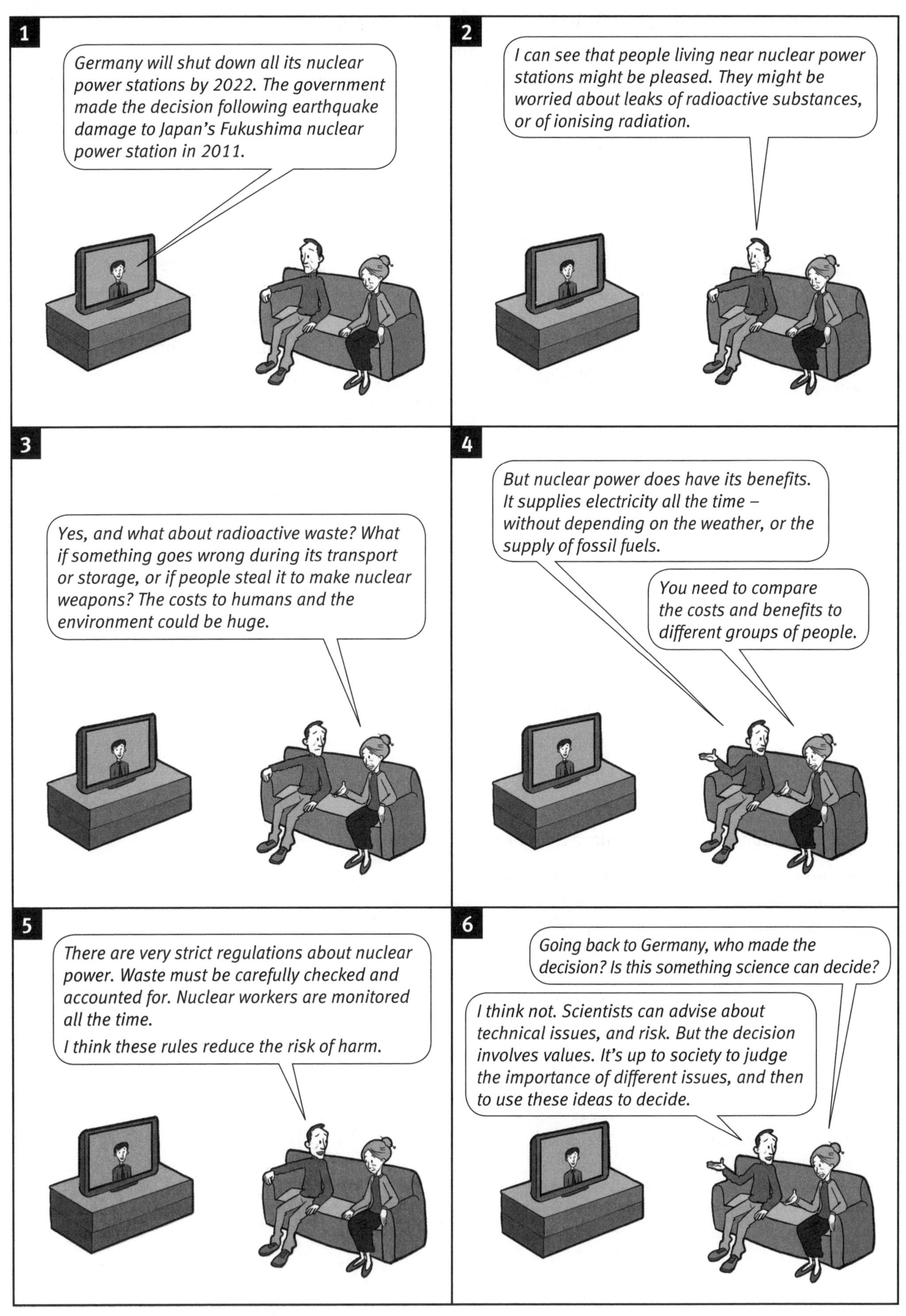

1
Germany will shut down all its nuclear power stations by 2022. The government made the decision following earthquake damage to Japan's Fukushima nuclear power station in 2011.
2
I can see that people living near nuclear power stations might be pleased. They might be worried about leaks of radioactive substances, or of ionising radiation.
3
Yes, and what about radioactive waste? What if something goes wrong during its transport or storage, or if people steal it to make nuclear weapons? The costs to humans and the environment could be huge.
4
But nuclear power does have its benefits. It supplies electricity all the time – without depending on the weather, or the supply of fossil fuels.
You need to compare the costs and benefits to different groups of people.
5
There are very strict regulations about nuclear power. Waste must be carefully checked and accounted for. Nuclear workers are monitored all the time.
I think these rules reduce the risk of harm.
6
Going back to Germany, who made the decision? Is this something science can decide?
I think not. Scientists can advise about technical issues, and risk. But the decision involves values. It's up to society to judge the importance of different issues, and then to use these ideas to decide.

Making decisions about science and technology

1 Imagine that a space organisation from the USA wants to build a new telescope outside the city of Addis Ababa, the capital city of Ethiopia. The telescope will collect visible and ultraviolet radiation from stars in our galaxy and beyond.

Different people make different claims about the telescope and its proposed site.

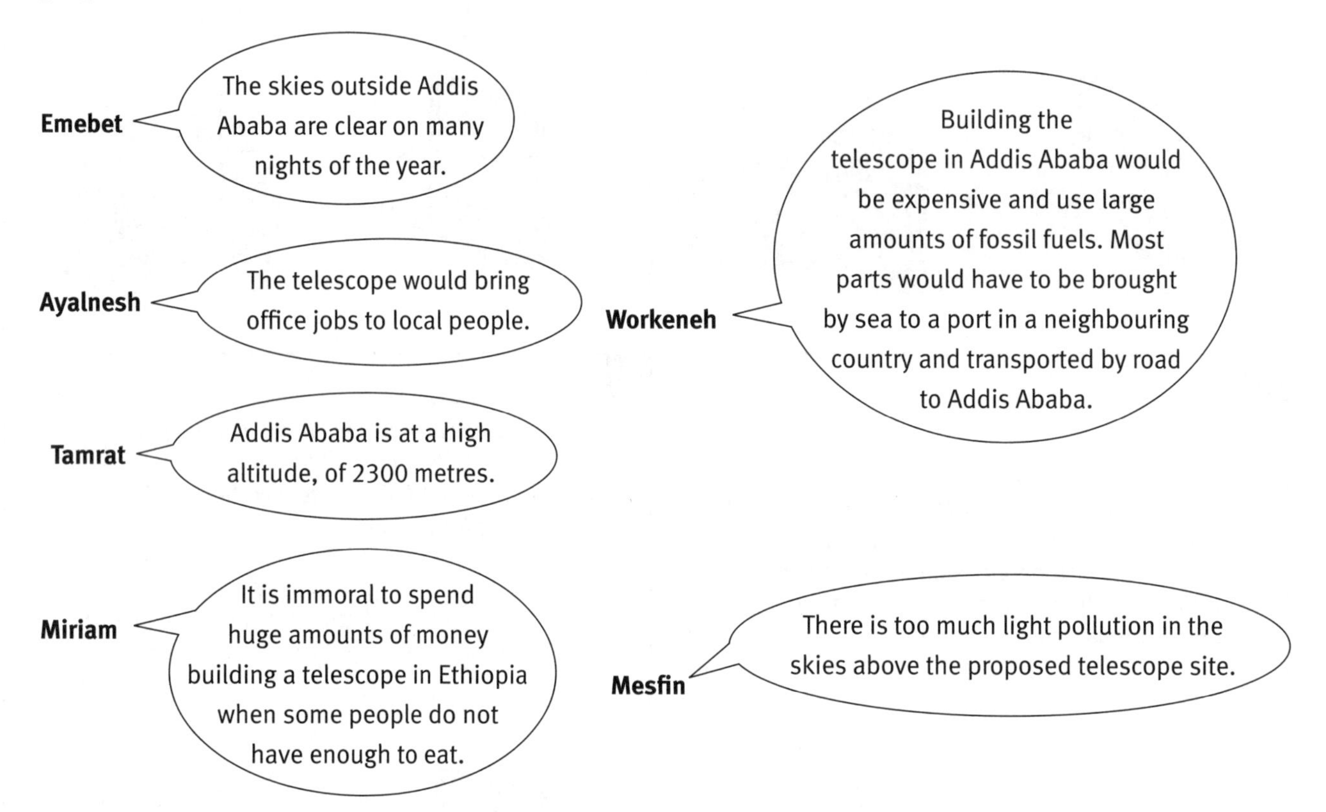

Emebet
The skies outside Addis Ababa are clear on many nights of the year.
Ayalnesh
The telescope would bring office jobs to local people.
Tamrat
Addis Ababa is at a high altitude, of 2300 metres.
Workeneh
Building the telescope in Addis Ababa would be expensive and use large amounts of fossil fuels. Most parts would have to be brought by sea to a port in a neighbouring country and transported by road to Addis Ababa.
Miriam
It is immoral to spend huge amounts of money building a telescope in Ethiopia when some people do not have enough to eat.
Mesfin
There is too much light pollution in the skies above the proposed telescope site.

Write the names of the people above in the correct box in the table below.

Each name may be used once, more than once, or not at all.

You can write one or more names in each box.

A person or people who identify...	Name or names
...a scientific reason for building the telescope in Addis Ababa.	
...a scientific reason against building the telescope in Addis Ababa.	
...a social impact of building the telescope in Addis Ababa.	
...an ethical issue.	
...an issue linked to sustainability.	
...a financial issue.	
...a claim that could be investigated scientifically.	

Making decisions about science and technology

1 a An electricity company needs to decide whether to build a new nuclear power station or a coal-fired power station.

The company asks people from five organisations for their opinions.

Freya Electricity from coal and nuclear power stations costs about the same.

Grace Coal-fired power stations produce acid rain. This damages trees.

Hari Nuclear power stations produce less carbon dioxide than coal-fired power stations. Carbon dioxide is a greenhouse gas.

Ian There is no risk-free way of disposing of nuclear waste. A nuclear waste leak could make land unsuitable for farming for hundreds of years.

Jasmine A nuclear accident could kill thousands. It is not right to put people at such a risk.

Kris If we build a nuclear power station, there will be more coal left for people in future.

i Write the names of two people whose opinions are linked to the idea of sustainability.

_________________ and _________________ [2]

ii Write the names of two people whose opinions show that they are concerned about the environment.

_________________ and _________________ [2]

iii Write the name of one person who is giving an ethical argument.

_________________ [1]

b Old nuclear power stations must be taken down. This process is called decommissioning.

Tick the boxes next to the jobs that are likely to be part of the work of an official organisation that regulates the decommissioning of nuclear power stations.

Making sure used fuel rods are disposed of safely. ☐

Deciding whether to build new nuclear or new gas-fired power stations ☐

Regularly checking storage sites where low-level and intermediate-level radioactive waste is stored. ☐

Deciding which countries to buy nuclear fuel from. ☐ [2]

Total [7]

2 Sustainability means using resources to meet the needs of people today without damaging the Earth or reducing the resources for people in future.

A power station in Bristol burns hospital waste to generate electricity.

a Suggest one way in which generating electricity from hospital waste is more sustainable than simply burning hospital waste.

____

____ [1]

b It is technically possible to use waste from all British hospitals to generate electricity.

However, waste from only a few hospitals is used in this way.

Suggest two reasons for this.

____

____

____ [2]

Total [3]

3 A British company wants to buy a large area of rainforest from the government of a poorer country.

It plans to clear the rainforest and use the land to grow oil palms.

It will transport palm oil to Britain.

In Britain, the palm oil will be burned to generate electricity.

- Summarise the benefits and costs the government might consider when deciding whether or not to sell the rainforest to the British company.
- On balance, do you think the government should sell the rainforest? Give reasons to support your decision.

✎ The quality of written communication will be assessed in your answer to this question.

Write your answer on separate paper or in your exercise book.

Total [6]

Exam tip

Before starting to write your answer to questions like question 3, make notes of the benefits and costs. Organise your notes, and decide on your decision. Then write your answer, based on these notes. Finally, check your answer and correct any mistakes of spelling, punctuation, or grammar.

Glossary

absolute zero Extrapolating from the behaviour of gases at different temperatures, the theoretically lowest possible temperature, –273°C. In practice, the lowest temperature achievable is about a degree above this.

absorb (radiation) The radiation that hits an object and is not reflected, or transmitted through it, is absorbed (for example, black paper absorbs light). Its energy makes the object get a little hotter.

absorption spectrum (of a star) Consists of dark lines superimposed on a continuous spectrum. It is created when light from the star passes through a cooler gas that absorbs photons of particular energies.

acceleration The rate of change of an object's velocity, that is, its change of velocity per second. In situations where the direction of motion is not important, the change of speed per second tells you the acceleration.

action at a distance An interaction between two objects that are not in contact, where each exerts a force on the other. Examples include two magnets, two electric charges, or two masses (for example, the Earth and the Moon).

activity The rate at which nuclei in a sample of radioactive material decay and give out alpha, beta, or gamma radiation.

actual risk Risk calculated from reliable data.

aerial A wire, or arrangement of wires, that emits radio waves when there is an alternating current in it, and in which an alternating current is induced by passing radio waves. So it acts as a source or a receiver of radio waves.

air resistance The force exerted on an object by the air, when it moves through it. Its direction is opposite to the direction in which the object is moving. It depends on the speed of the moving object (it increases as the object gets faster) and on its size and shape (as it is caused by the object having to move the air in front of it aside as it goes).

alpha radiation The least penetrating type of ionising radiation, produced by the nucleus of an atom in radioactive decay. A high-speed helium nucleus.

alternating current (a.c.) An electric current that reverses direction many times a second.

ammeter A meter that measures the size of an electric current in a circuit.

ampere (amp) The unit of electric current.

H **amplifier** A device for increasing the amplitude of an electrical signal. Used in radios and other audio equipment.

amplitude For a mechanical wave, the amplitude is the maximum distance that each point on the medium moves from its normal position as the wave passes. For an electromagnetic wave, it is the maximum value of the varying electric field (or magnetic field).

analogue signal Signal used in communications in which the amplitude can vary continuously.

angular magnification (of a refracting telescope) The ratio of the angle subtended by an object when seen through the telescope to the angle subtended by the same object when seen with the naked eye. It can be calculated as focal length of objective lens / focal length of eyepiece lens.

H **aperture (of a telescope)** The light-gathering area of the objective lens or mirror.

asteroid A dwarf rocky planet, generally orbiting the Sun between the orbits of Mars and Jupiter.

astrolabe An instrument used for locating and predicting the positions of the Sun, Moon, planets, and stars, as well as navigating and telling the time.

atmosphere The Earth's atmosphere is the layer of gases that surrounds the planet. It contains roughly 78% nitrogen and 21% oxygen, with trace amounts of other gases. The atmosphere protects life on Earth by absorbing ultraviolet solar radiation and reducing temperature extremes between day and night.

attract Pull towards.

average speed The distance moved by an object divided by the time taken for this to happen.

background radiation The low-level radiation, mostly from natural sources, that everyone is exposed to all the time, everywhere.

best estimate When measuring a variable, the value in which you have most confidence.

beta radiation One of several types of ionising radiation, produced by the nucleus of an atom in radioactive decay. More penetrating than alpha radiation but less penetrating than gamma radiation. A high-speed electron.

big bang An explosion of a single mass of material. This is currently the accepted scientific explanation for the start of the Universe.

black hole A mass so great that its gravity prevents anything escaping from it, including light. Some black holes are the collapsed remnants of massive stars.

H **calculated risk** Risk calculated from reliable data.

carbon cycle The human and natural processes that move carbon and carbon compounds continuously between the Earth, its oceans and atmosphere, and living things.

carrier A steady stream of radio waves produced by an RF oscillator in a radio to carry information.

cause When there is evidence that changes in a factor produce a particular outcome, then the factor is said to cause the outcome.

Cepheid variable A star whose brightness varies regularly, over a period of days.

chain reaction A process in which the products of one nuclear reaction cause further nuclear reactions to happen, so that more and more reactions occur and more and more product is formed. Depending on how this process is controlled, it can be used in nuclear weapons or the nuclear reactors in power stations.

charged Carrying an electric charge. Some objects (such as electrons and protons) are permanently charged. A plastic object can be charged by rubbing it. This transfers electrons to or from it.

climate Average weather in a region over many years.

coding (in communications) Converting information from one form to another, for example, changing an analogue signal into a digital one.

comet A rocky lump, held together by frozen gases and water, that orbits the Sun.

commutator An arrangement for changing the direction of the electric current through the coil of a motor every half turn. It consists of a ring divided into two halves (a split ring) with two contacts (called brushes) touching the two halves. It also allows the coil to turn continuously without the connecting wires getting tangled up.

compression A material is in compression when forces are trying to push it together and make it smaller.

conservation of energy The principle that the total amount of energy at the end of any process is always equal to the total amount of energy at the beginning – though it may now be stored in different ways and in different places.

constellation A group of stars that form a pattern in the night sky. Patterns recognised are cultural and historical, and are not based on the actual positions of the stars in space.

contamination (radioactive) Having a radioactive material inside the body, or having it on the skin or clothes.

control rod In a nuclear reactor, rods made of a special material that absorbs neutrons are raised and lowered to control the rate of fission reactions.

convective zone (of a star) The layer of a star above its radiative zone, where energy is transferred by convective currents in the plasma.

converging lens A lens that changes the direction of light striking it, bringing the light together at a point.

coolant In a nuclear reactor, the liquid or gas that circulates through the core and transfers heat to the boiler.

core The Earth's core is made mostly from iron, solid at the centre and liquid above.

correlation A link between two things. For example, if an outcome happens when a factor is present, but not when it is absent, there is a correlation between the outcome and the factor.

counter-force A force in the opposite direction to something's motion.

crust A rocky layer at the surface of the Earth, 10–40 km deep.

decoding In communications, converting information back into its original form, for example, changing a digital signal back into an analogue one.

decommissioning Taking a power station out of service at the end of its lifetime, dismantling it, and disposing of the waste safely.

detector Any device or instrument that shows the presence of radiation by absorbing it.

diffraction What happens when waves hit the edge of a barrier or pass through a gap in a barrier. They bend a little and spread into the region behind the barrier.

digital code A string of 0s and 1s that can be used to represent an analogue signal, and from which that signal can be reconstructed.

digital signals Signals used in communications in which the amplitude can take only one of two values, corresponding to the digits 0 and 1.

dioptre Unit of lens power, equivalent to a focal length of 1 metre.

direct current (d.c.) An electric current that stays in the same direction.

dispersion The splitting of white light into different colours (frequencies), for example, by a prism.

displacement The length and direction of the straight line from the initial position of an object to its position at a later time.

distance The length of the path along which an object has moved.

distance–time graph A graph showing the distance an object has moved along its path at each moment during its journey.

driving force The force pushing something forward, for example, a bicycle.

duration How long something happens for, for example, the length of time someone is exposed to radiation.

earthquake Event in which rocks break to allow tectonic plate movement, causing the ground to shake.

H **economic context** How money changes hands between businesses, government, and individuals.

efficiency The percentage of the energy supplied to a device that is transferred to the desired place, or in the desired way. For example, to find the efficiency of a kettle you would divide the gain in energy of the water by the work done on the kettle element by the electricity supply, and multiply by 100.

electric charge A fundamental property of matter. Electrons and protons are charged particles. Objects become charged when electrons are transferred to or from them, for example, by rubbing.

electric circuit A closed loop of conductors connected between the positive and negative terminals of a battery or power supply.

electric current A flow of charges around an electric circuit.

electric field A region where an electric charge experiences a force. There is an electric field around any electric charge.

electromagnetic induction The name of the process in which a potential difference (and hence often an electric current) is generated in a wire, when it is in a changing magnetic field.

electromagnetic spectrum The 'family' of electromagnetic waves of different frequencies and wavelengths.

electromagnetic wave A wave consisting of vibrating electric and magnetic fields, which can travel in a vacuum. Visible light is one example.

electron A tiny, negatively charged particle, which is part of an atom. Electrons are found outside the nucleus. Electrons have negligible mass and one unit of charge.

electrostatic attraction The force of attraction between objects with opposite electric charges. A positive ion, for example, attracts a negative ion.

emission spectrum (of an element) The electromagnetic frequencies emitted by an excited atom as electron energy levels fall.

erosion The movement of solids at the Earth's surface (for example, soil, mud, rock) caused by wind, water, ice, and gravity, or living organisms.

ethics A set of principles that may show how to behave in a situation.

exoplanet The planet of any star other than the Sun.

extended object An astronomical object made up of many points, for example, the Moon or a galaxy. By contrast, a star is a single point.

eyepiece lens (of an optical telescope) The lens nearer the eye, which will have a higher power. Often called a telescope 'eyepiece'.

factor A variable that changes and may affect something else.

focal length The distance from the optical centre of a lens or mirror to its focus.

focus (of a lens or mirror) The point at which rays arriving parallel to its principal axis cross each other. Also called the 'focal point'.

focusing Adjusting the distance between lenses, or between the eyepiece lens and a photographic plate (or CCD), to obtain a sharp image of the object.

force A push or a pull experienced by an object when it interacts with another. A force is needed to change the motion of an object.

fossil The stony remains of an animal or plant that lived millions of years ago, or an imprint it has made (for example, a footprint) in a surface.

fossil fuel Natural gas, oil, or coal.

friction The force exerted on an object due to the interaction between it and another object that it is sliding over. It is caused by the roughness of both surfaces at a microscopic level.

fuel rod A container for nuclear fuel, which enables fuel to be inserted into, and removed from, a nuclear reactor while it is operating.

galaxy A collection of thousands of millions of stars held together by gravity.

gamma radiation (gamma rays) The most penetrating type of ionising radiation, produced by the nucleus of an atom in radioactive decay. The most energetic part of the electromagnetic spectrum.

generator A device that uses motion to generate electricity. It consists of a coil that is rotated in a magnetic field. This produces a potential difference across the ends of the coil, which can then be used to provide an electric current.

geothermal power station A power station that uses hot underground rocks to heat water to produce steam to drive turbines.

globular cluster A cluster of hundreds of thousands of old stars.

gravitational potential energy The energy stored when an object is raised to a higher point in the Earth's gravitational field.

greenhouse effect The atmosphere absorbs infrared radiation from the Earth's surface and radiates some of it back to the surface, making it warmer than it would otherwise be.

greenhouse gas Gases that contribute to the greenhouse effect. Includes carbon dioxide, methane, and water vapour.

half-life The time taken for the amount of a radioactive element in a sample to fall to half its original value.

high-level waste A category of nuclear waste that is highly radioactive and hot. Produced in nuclear reactors and nuclear weapons processing.

H **Hubble constant** The ratio of the speed of recession of galaxies to their distance, with a value of about 72 km/s per Mpc.

in parallel A way of connecting electric components that makes a branch (or branches) in the circuit so that charges can flow around more than one loop.

in series A way of connecting electric components so that they are all in a single loop. The charges pass through them all in turn.

infrared Electromagnetic waves with a frequency lower than that of visible light, beyond the red end of the visible spectrum.

instantaneous speed The speed of an object at a particular instant. In practice, its average speed over a very short time interval.

intensity (of light in a star's spectrum) The amount of a star's energy gathered by a telescope every second, per unit area of its aperture.

interaction What happens when two objects collide, or influence each other at a distance. When two objects interact, each experiences a force.

interaction pair Two forces that arise from the same interaction. They are equal in size and opposite in direction, and each acts on a different object.

interference This happens when two waves meet. If the waves have the same frequency, an interference pattern is formed. In some places, crests add to crests, forming bigger crests; in other places, crests and troughs cancel each other out.

intermediate-level waste A category of nuclear waste that is generally short-lived but requires some shielding to protect living organisms, for example, contaminated materials that result from decommissioning a nuclear reactor.

H **intrinsic brightness (of a star)** A measure of the light that would reach a telescope if a star were at a standard distance from the Earth.

inverted image An image that is upside down compared to the object.

ionising radiation Radiation with photons of sufficient energy to remove electrons from atoms in its path. Ionising radiation, such as ultraviolet, X-rays, and gamma rays, can damage living cells.

irradiation Being exposed to radiation from an external source.

isotope Atoms of the same element that have different mass numbers because they have difference numbers of neutrons in the nucleus.

Kelvin scale A scale of temperature in which 0 K is absolute zero, and the triple point of water (where solid, liquid, and gas phases co-exist) is 273 K.

kinetic energy The energy that something has owing to its motion.

kinetic model of matter The idea that a gas consists of particles (atoms or molecules) that move around freely, colliding with each other and with the walls of any container, with most of the volume of gas being empty space.

light pollution Light created by humans, for example, street lighting, that prevents city dwellers from seeing more than a few bright stars. It also causes problems for astronomers.

light-dependent resistor (LDR) An electric circuit component whose resistance varies depending on the brightness of light falling on it.

light-year The distance travelled by light in a year.

longitudinal wave A wave in which the particles of the medium vibrate in the same direction as the wave is travelling. Sound is an example.

low-level waste A category of nuclear waste that contains small amounts of short-lived radioactivity, for example, paper, rags, tools, clothing, and filters from hospitals and industry.

luminosity (of a star) The amount of energy radiated into space every second. This can be measured in watts, but astronomers usually compare a star's luminosity to the Sun's luminosity.

lunar eclipse This occurs when the Earth comes between the Moon and the Sun, and totally or partially covers the Moon in the Earth's shadow as seen from the Earth's surface.

magnetic field The region around a magnet, or a wire carrying an electric current, in which magnetic effects can be detected. For example, another small magnet in this region will experience a force and tend to move.

magnification (of an optical instrument) The process of making something appear closer than it really is.

mantle A thick layer of rock beneath the Earth's crust, which extends about halfway down to the Earth's centre.

mean value A type of average, found by adding up a set of measurements and then dividing by the number of measurements.

megaparsec (Mpc) A million parsecs.

microwaves Radio waves of the highest frequency (shortest wavelength), used for mobile phones and satellite TV.

Milky Way The galaxy in which the Sun and its planets including Earth are located. It is seen from the Earth as an irregular, faintly luminous band across the night sky.

H **modulate** To vary the amplitude or frequency of carrier waves so that they carry information.

momentum (plural momenta) A property of any moving object. Equal to mass multiplied by velocity.

motor A device that uses an electric current to produce continuous motion.

mountain chain A group of mountains that extend along a line, often hundreds or even thousands of kilometres. Generally caused by the movement of tectonic plates.

nanometre A unit used for microscopic measurements. 1 nm = 0.001 µm = 0.000 001 mm.

negative A label used to name one type of charge or one terminal of a battery. It is the opposite of positive.

neutron star The collapsed remnant of a massive star, after a supernova explosion. Made almost entirely of neutrons, they are extremely dense.

noise Unwanted electrical signals that get added on to radio waves during transmission, causing additional modulation. Sometimes called 'interference'.

non-ionising radiation Radiation with photons that do not have enough energy to ionise molecules.

normal An imaginary line drawn at right angles to the point at which a ray strikes the boundary between one medium and another. Used to define the angle of the ray that strikes or emerges from the boundary.

nuclear fission The process in which a nucleus of uranium-235 breaks apart, releasing energy, when it absorbs a neutron.

nuclear fuel In a nuclear reactor, each uranium atom in a fuel rod undergoes fission and releases energy when hit by a neutron.

nuclear fusion The process in which two small nuclei combine to form a larger one, releasing energy. An example is hydrogen combining to form helium. This happens in stars, including the Sun.

objective lens (of an optical telescope) The lens nearer the object, which will have a lower power. Often called a telescope 'objective'.

observed brightness (of a star) A measure of the light reaching a telescope from a star.

oceanic ridge A line of underwater mountains in an ocean, where new seafloor constantly forms.

ohm The unit of electrical resistance. Symbol Ω.

Ohm's law The result that the current, *I*, through a resistor, *R*, is proportional to the voltage, *V*, across the resistor, provided its temperature remains the same. Ohm's law does not apply to all conductors.

optical fibre Thin glass fibre down which a light beam can travel. The beam is reflected at the sides by total internal reflection, so very little escapes. Used in modern communications, for example, to link computers in a building into a network.

outcome A variable that changes as a result of something else changing.

outlier A measured result that seems very different to other repeat measurements, or from the value you would expect, which you therefore strongly suspect is wrong.

ozone layer A thin layer in the atmosphere, about 30 km up, where oxygen is in the form of ozone molecules. The ozone layer absorbs ultraviolet radiation from sunlight.

parallax The apparent shift of an object against a more distant background, as the position of the observer changes. The further away an object is, the less it appears to shift. This can be used to measure how far away an object is, for example, to measure the distance to stars.

parallax angle When observed at an interval of six months, a star will appear to move against the background of much more distant stars. Half of its apparent angular motion is called its parallax angle.

parsec (pc) A unit of astronomical distance, defined as the distance of a star that has a parallax angle of one arcsecond. Equivalent to 3.1×10^{13} km.

peak frequency The frequency with the greatest intensity.

peer review The process whereby scientists who are experts in their field critically evaluate a scientific paper or idea before and after publication.

penumbra An area of partial darkness in a shadow, for example, places in the Moon's path where the Earth only partially blocks off sunlight. Some sunlight still reaches these places because the Sun has such a large diameter.

perceived risk The level of risk that people think is attached to an activity, not based on data.

phases (of the Moon) Changing appearance, due to the relative positions of the Earth, Sun, and Moon.

photon Tiny 'packet' of electromagnetic radiation. All electromagnetic waves are emitted and absorbed as photons. The energy of a photon is proportional to the frequency of the radiation.

photosphere The visible surface of a star, which emits electromagnetic radiation.

planet A very large, spherical object that orbits the Sun, or other star.

positive A label used to name one type of charge, or one terminal of a battery. It is the opposite of negative.

potential difference (p.d.) The difference in potential energy (for each unit of charge flowing) between any two points in an electric circuit. Also called voltage.

power In an electric circuit, the rate at which work is done by the battery or power supply on the components in a circuit. Power is equal to current × voltage.

pressure (of a gas) The force a gas exerts per unit area on the walls of its container.

primary energy source A source of energy not derived from any other energy source, for example, fossil fuels or uranium.

principal axis An imaginary line perpendicular to the centre of a lens or mirror surface.

proportional Two variables are proportional if there is a constant ratio between them.

proton A positively charged particle found in the nucleus of atoms. The relative mass of a proton is 1 and it has one unit of charge.

protostar The early stages in the formation of a new star, before the onset of nuclear fusion in the core.

P-wave A seismic wave through the Earth, produced during an earthquake.

radiation A flow of information and energy from a source. Light and infrared are examples. Radiation spreads out from its source, and may be absorbed or reflected by objects in its path. It may also go (be transmitted) through them.

radiation dose A measure, in millisieverts, of the possible harm done to your body, which takes into account both the amount and type of radiation you are exposed to.

radiative zone (of a star) The layer of a star surrounding its core, where energy is transferred by photons to the convective zone.

radio waves Electromagnetic waves of a much lower frequency than visible light. They can be made to carry signals and are widely used for communications.

radioactive Used to describe a material, atom, or element that produces alpha, beta, or gamma radiation.

radioactive dating Estimating the age of an object such as a rock by measuring its radioactivity. Activity falls with time, in a way that is well understood.

radioactive decay The spontaneous change in an unstable element, giving out alpha, beta, or gamma radiation. Alpha and beta emission result in a new element.

radiotherapy Using radiation to treat a patient.

random Of no predictable pattern.

range The difference between the highest and the lowest of a set of data.

ray diagram A way of representing how a lens or telescope affects the light that it gathers, by drawing the rays (which can be thought of as very narrow beams of light) as straight lines.

reaction (of a surface) The force exerted by a hard surface on an object that presses on it.

real difference One way of deciding if there is a real difference between two values is to look at the mean values and the ranges. The difference between two mean values is real if their ranges do not overlap.

reflection What happens when a wave hits a barrier and bounces back off it. If you draw a line at right angles to the barrier, the reflected wave has the same angle to this line as the incoming wave. For example, light is reflected by a mirror.

reflector A telescope that has a mirror as its objective. Also called a reflecting telescope.

refraction Waves change their wavelength if they travel from one medium to another in which their speed is different. For example, when travelling into shallower water, waves have a smaller wavelength as they slow down.

refractor A telescope that has a lens as its objective, rather than a mirror.

renewable energy source Resources that can be used to generate electricity without being used up, such as the wind, tides, and sunlight.

repeatable A quality of a measurement that gives the same result when repeated under the same conditions.

repel Push apart.

reproducible A quality of a measurement that gives the same results when carried out under different conditions, for example, by different people or using different equipment or methods.

resistance The resistance of a component in an electric circuit indicates how easy or difficult it is to move charges through it.

resolving power The ability of a telescope to measure the angular separation of different points in the object that is being viewed. Resolving power is limited by diffraction of the electromagnetic waves being collected.

resultant force The sum, taking their directions into account, of all the forces acting on an object.

retrograde motion An apparent reversal in a planet's usual direction of motion, as seen from the Earth against the background of fixed stars. This happens periodically with all planets beyond the Earth's orbit.

risk The probability of an outcome that is seen as undesirable, associated with some behaviour or process.

rock cycle Continuing changes in rock material, caused by processes such as erosion, sedimentation, compression, and heating.

sampling In the context of physics, measuring the amplitude of an analogue signal many times a second in order to convert it into a digital signal.

seafloor spreading The process of forming new ocean floor at oceanic ridges.

secondary energy source Energy in a form that can be distributed easily but is manufactured by using a raw energy resource such as a fossil fuel or wind. Examples of secondary energy sources are electricity, hot water used in heating systems, and steam.

selective absorption Some materials absorb some forms of electromagnetic radiation but not others. For example, glass absorbs infrared but is transparent to visible light.

sidereal day The time taken for the Earth to rotate 360°: 23 hours and 56 minutes.

signal Information carried through a communication system, for example, by an electromagnetic wave with variations in its amplitude or frequency, or being rapidly switched on an off.

slope The slope of a graph is a measure of its steepness. It is calculated by choosing two points on the graph and calculating: the change in the value of the y-axis variable/ the change in the value of the x-axis variable. If the graph is a straight line, you can use any two points on it. If it is curved, you can estimate the slope at any chosen point, using two points on the graph close to this point.

social context The situation of people's lives.

solar day The time taken for the Earth to rotate so that it fully faces the Sun again: exactly 24 hours.

solar eclipse When the Moon comes between the Earth and the Sun, and totally or partially blocks the view of the Sun as seen from the Earth's surface.

Solar System The Sun and objects that orbit around it – planets and their moons, comets, and asteroids.

source An object that produces radiation.

spectrometer An instrument that divides a beam of light into a spectrum and enables the relative brightness of each part of the spectrum to be measured.

spectrum One example is the continuous band of colours, from violet to red, produced by shining white light through a prism. Passing light from a flame test through a prism produces a line spectrum.

speed of light 300 000 kilometres per second – the speed of all electromagnetic waves in a vacuum.

speed of recession The speed at which a galaxy is moving away from us.

star life cycle All stars have a beginning and an end. Physical processes in a star change throughout its life, affecting its appearance.

static electricity Electric charge that is not moving around a circuit but has built up on an object such as a comb or a rubbed balloon.

strong (nuclear) force A fundamental force of nature that acts inside atomic nuclei.

Sun The star nearest Earth. Fusion of hydrogen in the Sun releases energy, which makes life on Earth possible.

H **S-wave** A transverse seismic wave through the Earth, produced during an earthquake.

supernova A dying star that explodes violently, producing an extremely bright astronomical object for weeks or months.

sustainability Using resources and the environment to meet the needs of people today without damaging Earth or reducing the resources for people in future.

sustainable Meeting the needs of people today without damaging the Earth for future generations.

tectonic plate Giant slabs of rock (about 12, comprising crust and upper mantle) that make up the Earth's outer layer.

telescope An instrument that gathers electromagnetic radiation, to form an image or to map data, from astronomical objects such as stars and galaxies. It makes things visible that cannot be seen with the naked eye.

tension A material is in tension when forces are trying to stretch it or pull it apart.

theory A scientific explanation that is generally accepted by the scientific community.

thermistor An electric circuit component whose resistance changes markedly with its temperature. It can therefore be used to measure temperature.

transformer An electrical device consisting of two coils of wire wound on an iron core. An alternating current in one coil causes an ever-changing magnetic field that induces an alternating current in the other. Used to 'step' voltage up or down to the level required.

transmitted (transmit) When radiation hits an object, it may go through it. It is said to be transmitted through it. We also say that a radio aerial transmits a signal. In this case, transmits means 'emits' or 'sends out'.

transverse wave A wave in which the particles of the medium vibrate at right angles to the direction in which the wave is travelling. Water waves are an example.

ultraviolet (UV) Electromagnetic waves with frequencies higher than those of visible light, beyond the violet end of the visible spectrum.

umbra An area of total darkness in a shadow. For example, places in the Moon's path where the Earth completely blocks off sunlight.

uncertain Describes measurements in a situation where scientists know that they may not have recorded the true value.

Universe All things (including the Earth and everything else in space).

unstable The nucleus in radioactive isotopes is not stable. It is liable to change, emitting one of several types of radiation. If it emits alpha or beta radiation, a new element is formed.

velocity The speed of an object in a given direction. Unlike speed, which only has a size, velocity also has a direction.

velocity–time graph A useful way of summarising the motion of an object by showing its velocity at every instant during its journey.

vibration Moving rapidly and repeatedly back and forth.

volcano A vent in the Earth's surface that erupts magma, gases, and solids.

voltage The voltage marked on a battery or power supply is a measure of the 'push' it exerts on charges in an electric circuit. The 'voltage' between two points in a circuit means the 'potential difference' between these points.

voltmeter An instrument for measuring the potential difference (which is often called the 'voltage') between two points in an electric circuit.

wave speed The speed at which waves move through a medium.

wavelength The distance between one wave crest (or wave trough) and the next.

work Work is done whenever a force makes something move. The amount of work is force multiplied by distance moved in the direction of the force. This is equal to the amount of energy transferred.

X-ray Electromagnetic wave with high frequency, well above that of visible light.

Answers

P1 Workout

1 a C b A c D d A

2 a 12 700 b Four thousand million
c 10 d Five thousand million
e 14 thousand million

3 G, F, E, A, H, I, D, B, C

4 a Comet: big lump of ice and dust that rushes past the Sun and then returns to the outer Solar System; asteroid: lump of rock that is usually smaller than a comet, with an almost circular orbit.
b Moon: orbits a planet, and is smaller than the planet it orbits; planet: orbits the Sun and is bigger than its moons.
c Star: ball of hot gas; galaxy: made up of thousands of millions of stars.

5 Vibrations at right angles – transverse – S-waves and water waves
Vibrations in same direction – longitudinal – P-waves and sound waves

6

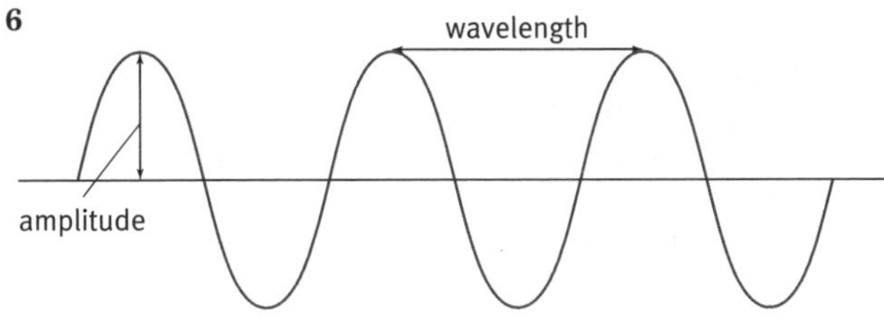

7 a 333 m/s b 10 km/s

P1 Quickfire

1 Moving out from the centre: core, mantle, crust
2 Objects that emit light: **b**, **c**
3 Brightness, pollution, uncertainties, assumptions.
4 a A travelling vibration that transfers energy from place to place without transferring matter
b The number of waves that pass any point each second
c The distance between two corresponding points on adjacent cycles
d The distance from the maximum displacement to the undisturbed position
e The distance light travels through a vacuum in one year
5 a 2050 km b 2050 km c 20 400 m
6 C, A, D, B, E
7 Shapes of continents seem to fit together; matching fossils in eastern South America and western Africa; matching rock types in eastern South America and western Africa
8 Wegener was an outsider to the community of geologists; the movement of the continents was not detectable with the measuring instruments of the time; the idea seemed too big from the limited evidence available.
9 a 7.5 km/s b 9 km/s
10 a About 2060 km
b The speed of the S-waves is 0 between a depth of about 2060 km and 5000 km. This is evidence that the material that makes up the outer core is liquid, since S-waves cannot travel through liquids.
11 a Redshift – the shifting of light emitted by distant galaxies towards the red end of the spectrum – is evidence for distant galaxies moving away from us.
b Galaxies that are further away from us moving faster than those that are closer is evidence that space is expanding.
c The pattern of stripes in the rocks formed at oceanic ridges is evidence that the Earth's magnetic field has changed direction several times in the past.
12 See the diagram of the rock cycle in the Factbank.

P1 GCSE-style questions

1 a Galileo saw Ganymede through a telescope.
Gan Dej saw an object close to Jupiter with a telescope.
Ganymede is a moon.
b When Gan Dej made his observations, the skies would have been very dark. The apparent magnitude of Ganymede is smaller than that of the faintest object in Space that can be seen in a very dark sky without a telescope, so it is possible for Gan Dej to have seen Ganymede.
c i Ganymede orbits a planet, not a star.
ii The mass of Pluto is less than the total mass of the other objects that cross its orbit.
iii Xena orbits the Sun.

2 a From centre: core, mantle, crust b 1C; 2D; 3B; 4A

3 a P-waves travel faster. They arrive at the seismometer first.
b i

Seismograph	S-P time interval (s)
A	approximately 52
B	approximately 74
C	approximately 20

ii Seismograph C was recorded closest to the earthquake. The S-P interval is smallest.
c i Distance = speed × time
Distance = 5 km/s × 60 s
Distance = 300 km
ii Owlton and Badgerbridge

P2 Workout

1 The satellite both absorbs and emits radiation. The air transmits radiation. The transmitter is a source of radiation. The satellite dish is a detector. It absorbs radiation. The energy deposited here by a beam of radiation depends on the number of photons and the energy of one photon. The hill reflects radiation. The energy that arrives at a square metre surface each second is the intensity of the radiation.

2 1 radio, 2 infrared, 3 information, 4 optical, 5 receiver, 6 AM; 7 analogue; 8 digital; 9 interference, 10 decode; 11 noise

3 1 Radio waves, no damage. 2 Microwaves; heat up cells and damage them, protect by shutting microwave oven door. 3 Microwaves: may heat up cells and damage them in young children; protect children by making them use hands free or texting instead of phoning. 4 X-rays; ionise atoms or molecules and so damage DNA of living cells, which may lead to cancer or cell death; protect by using only when necessary and dentist to be out of the room, or behind lead shield. 5 UV; damages cells, leading to cell death or cancer; protect by covering up with clothes, wearing sunscreen, keeping in the shade. 6 Gamma rays; in nuclear power station; cause cancer and cell death; keep out! 7 Light waves from TV; no damage.

P2 Quickfire

1 Ionising radiations: a, c, e; radiations that cause a heating effect only: b, d, f
2 G, D, B, C, F, A, E
3 a C b O c B
d O e C f C
4 True statements: b, c
Corrected version of false statement:
a The higher the frequency of an electromagnetic radiation, the more energy is transferred by each photon.

5 Microwaves, radio or TV, radio or TV, absorbed, visible light or infrared, visible light or infrared, absorbed
6 X: it is made up of the greatest amount of stored information.
7 In order, from top of left column: digital signal without noise; analogue signal with noise; analogue signal without noise; digital signal with noise.
8 a Respiration and volcanic activity.
  b One from: photosynthesis, dissolving, sedimentation/forming carbonaceous rocks.
  c Humans have burned more fossil fuels; forests have been cleared for farmland.
  d Methane and water vapour.
  e Climate change, meaning some food crops will no longer grow in some places; ice melting and seawater expanding as it warms up, causing rising sea levels and flooding of low lying land; more extreme weather conditions.
9 a Water
  b Sunscreen and clothing
  c Lead and other dense materials
10 The energy arriving at a square metre of a surface each second
11 Because it reaches an ever-increasing surface area, and because some of the radiation is absorbed by the medium it is travelling through
12 Ozone O_3; oxygen O_2. When an ozone molecule absorbs ultraviolet radiation, the molecule may break down.
13 The principal frequency of the radiation emitted by an object is the frequency that is emitted with the highest intensity. The Sun has a higher principal frequency than the Earth.
14 Higher temperatures cause more convection in the atmosphere, and more evaporation of water from oceans and land.

P2 GCSE-style questions

1 a i Two from: Combustion, respiration, volcanic activity, decomposition
    ii Two from: Photosynthesis, dissolving in water, sedimentation/forming carbonaceous rocks
  b i Rising sea levels; changing climates
    ii Methane, water vapour
2 a Transmitted
  b Absorbs; if the intensity of the light that comes out of the brain is less than expected, the brain might be bleeding. This suggests that blood absorbs light, so the intensity of transmitted light decreases.
  c In one second, a different number of photons arrives at each detector.
  d i One of: X-rays damage living cells and can cause cancer; X-rays will be absorbed by the bone of the skull, so will give no useful information.
    ii Microwaves have a heating effect, so will heat up brain tissue causing great damage
3 5/6 marks
Answer clearly describes all the trends shown by the graphs, and refers to data from them.
and identifies correlations shown by pairs of graphs.
and indicates that correlations are not necessarily causal.
All information in the answer is relevant, clear, organised and presented in a structured and coherent format. Specialist terms are used appropriately. Few, if any, errors in grammar, punctuation and spelling.
3/4 marks
Answer identifies trends shown by some of the graphs, and refers to data from some of them.
and identifies two or three correlations shown by the graphs.
Most of the information is relevant and presented in a structured and coherent format. Specialist terms are usually used correctly. There are occasional errors in grammar, punctuation and spelling.
1/2 marks
Answer indentifies trends shown by some of the graphs, but does not refer to data from them.
or identifies one or two correlations shown by the graph.
and may incorrectly state or imply that the correlations are necessarily causal.
There may be limited use of specialist terms. Errors of grammar, punctuation and spelling prevent communication of the science. Answer includes 1 or 2 points of those listed below.
0 marks
Insufficient or irrelevant science. Answer not worthy of credit.
Relevant points include:
- Graph A shows that average temperature has increased since 1880, with the most rapid increases being since about 1950.
- Graphs B and E show that the concentrations of methane and carbon dioxide in the atmosphere have increased over time.
- Graphs A, B, and E show that here are correlations between average temperature and the concentrations of methane and carbon dioxide.
- The correlation does not prove that increasing concentrations of methane and carbon dioxide are the cause of global warming.
- Graph C shows that sea levels have increased over time.
- Graph D shows that sea pH has decreased over time.
- Graphs A and C show that there is a correlation between average temperature and sea levels.
- The correlation does not prove that increasing average temperatures increases sea levels.
- Graphs B and D show that there is a correlation between carbon dioxide concentration and sea pH. The correlation does not prove that increasing carbon dioxide concentration in the atmosphere causes decreased sea pH.

4 The intensity of the radiation arriving at Helen's phone is less *or* the distance between Helen and the source is greater.

P3 Workout

1 1 B or C or G; 2 A; 3 D; 4 F; 5 B or C or G; 6 B or C or G; 7 E
2 –
3 Person on left: A, C, D, E; person on right: B, D, E, F

P3 Quickfire

1 **b, c, d, e, f**
2 Voltage – volt; energy – kilowatt-hour, joule; current – amp; time – hour, second; power – watt
3

	Biofuels	Gas	Solar
It is a primary energy source.	√	√	√
It is a renewable energy source.	√		√
When electricity is generated from it, CO_2 is produced.	√	√	
The generation of electricity from this source is weather dependent.			√
When electricity is generated from it, no waste is produced.			√

4 A, F, B, D, C, G, E

5

Device	Power rating (W)	Power rating (kW)	Time it is on for	Energy transferred (kWh)
computer	250	0.250	2 hour	0.5
kettle	1800	1.800	3 min	0.09
toaster	1200	1.200	5 min	0.1
phone charger	20	0.02	2 hour	0.04

6 Switch off appliances, use energy efficient appliances, walk instead of going by car. *Other points are acceptable here.*

7 Provide more efficient public transport, generating electricity more efficiently, making laws about energy efficiency of new homes and workplaces

8 **a** 23.75p **b** 10p

9 **a** 17% **b** 27 kJ

10 Because energy is conserved, so the total amount of energy must remain the same.

11 The power of a device or appliance is the rate at which it transfers energy.

12 A country needs a mix of energy sources to ensure continuity of supply.

P3 GCSE-style questions

1 **a** **i** 120 × 2 MW gains I mark 240 gains the second mark.

ii 5760 MWh 240 MW × 24h gains I mark. 5760 MWh gains the seond mark.

b 1 163 077 people

c **i** To follow the Sun so as to maximise the energy that it receives

ii 15%

d There may be no more places in which hydroelectric power stations can be built; it is wise to have a mix of energy sources to ensure continuity of supply.

e **i** The energy is renewable (or any other sensible answer).

ii They were worried about hazards to shipping, or possible damage to wildlife.

2 5/6 marks: answer clearly identifies several advantages **and** disadvantages of replacing fossil fuels with biofuels **and** comes to a judgement about which is the better choice **and** gives clear reasons for the judgement, linked to the advantages and disadvantages given.

All information in the answer is relevant, clear, organised and presented in a structured and coherent format. Specialist terms are used appropriately. Few, if any, errors in grammar, punctuation, and spelling.

3/4 marks: answer identifies some advantages **and** disadvantages of replacing fossil fuels with biofuels **and** gives to a judgement about which is the better choice, **but** does not give reasons for the judgement.

Most of the information is relevant and presented in a structured and coherent format. Specialist terms are usually used correctly. There are occasional errors in grammar, punctuation, and spelling.

1/2 marks: answer identifies a few advantages **or** disadvantages of replacing fossil fuels with biofuels **or** gives a judgement about which is the better choice **but** does not give reasons for the judgement.

There may be limited use of specialist terms. Errors of grammar, punctuation, and spelling prevent communication of the science. Answer includes 1 or 2 points of those listed below.

0 marks: insufficient or irrelevant science. Answer not worthy of credit.

Relevant points include:

- Fossil fuels and biofuels produce carbon dioxide (a greenhouse gas) when they burn.
- Fossil fuels and biofuels may produce other gaseous pollutants, and particulates, when they burn.
- Fossil fuels are not renewable.
- Biofuels are renewable.
- Plants from which biofuels are made take in carbon dioxide from the atmosphere as they grow.
- Plants from which biofuels are made may be grown on land that could be used to grow food.
- An evaluation referring to the points above – is the government's idea to replace fossil fuels with biofuels a good one?

3 **a** A furnace; B steam; C generator; D water; E turbine; F transformer

b 25%

c It produces huge amounts of carbon dioxide

d For: the nuclear power station does not produce carbon dioxide when it is being used. Against: the nuclear power station relies on fuel that may need to be imported; there are radiation hazards associated with nuclear power stations. *Other points are acceptable here.*

P4 Workout

1 **a** F

b T

c F

d T

e F

f T

2 Rope 10 N to right; tricycle 120 N to right; trolley no resultant force

3 Left picture: C; middle picture: B; right picture: A

4 In order along the curve: B, A, C, E, D, F

5 1 interaction; 2 kinetic; 3 acceleration; 4 resultant; 5 friction; 6 average; 7 reaction; 8 potential; 9 driving; 10 momentum; 11 negative

P4 Quickfire

1 Average, time, short, gravity, upwards, constant

2 **a** T

b T

c T

d F

e T

f T

g F

3 **a** Greater

b Ali, the weight

c Less, heating

4 **a** 200 m/min

b 40 m/s

c 18 m/s

d 4 cm/s

5 **a** 5 m/s^2

b 2 m/s^2

6 **a** 88 000 kg m/s

b 304 kg m/s

c 13.5 kg m/s

7 15 000 Ns

8 **a** 30 000 J

b 1 250 000 J

c 58 J

9 A jet engine draws in air at the front and pushes it out at the back. An equal and opposite force pushes the engine forward. This is the driving force.
10 In order along the curve: A, B, D, F, E, C

P4 GCSE-style questions

1 a i All statements are true except the first one.
ii 1500 J
iii 1500 J
b 300 J
2 5/6 marks: answer explains in detail why the driving force needed by lorry B is less than the driving force needed by lorry A **and** the explanation is logical and coherent.
All information in the answer is relevant, clear, organised, and presented in a structured and coherent format. Specialist terms are used appropriately. There are few, if any, errors in grammar, punctuation, and spelling.
3/4 marks: answer explains why the driving force needed by lorry B is less than the driving force needed by lorry A **but** the answer lacks detail **or** the explanation lacks clarity and coherence.
Most of the information is relevant and presented in a structured and coherent format. Specialist terms are usually used correctly. There are occasional errors in grammar, punctuation, and spelling.
1/2 marks: answer gives some reasons to explain the difference in driving force **and** the explanation lacks detail **and** the answer lacks clarity and coherence.
There may be limited use of specialist terms. Errors of grammar, punctuation, and spelling prevent communication of the science. Answer includes 1 or 2 points of those listed below.
0 marks: insufficient or irrelevant science. Answer not worthy of credit.
Relevant points to include:
- For the lorry to travel in a straight line at a steady speed, the driving force needs to be equal to the counter force.
- The counter force is made up of friction and air resistance.
- The frictional force between the road and the lorry is the same for each lorry.
- Lorry B experiences less air resistance.
- So the counter force on lorry B is less.
- So the driving force needed by lorry B is less.

3 a i D to E
ii Stationary from B to C; moving at a steady speed from C to D – this is the fastest part of the fire engine's journey.
iii 1 km/min
b 12 m/s
4 a F = cat and G = horse
b Animal A top speed = 3 m/s; animal B top speed = 6 m/s
c Correct statements: the top speed of animal B is twice the top speed of animal A; animal C is unlikely to be a pig.
5 a

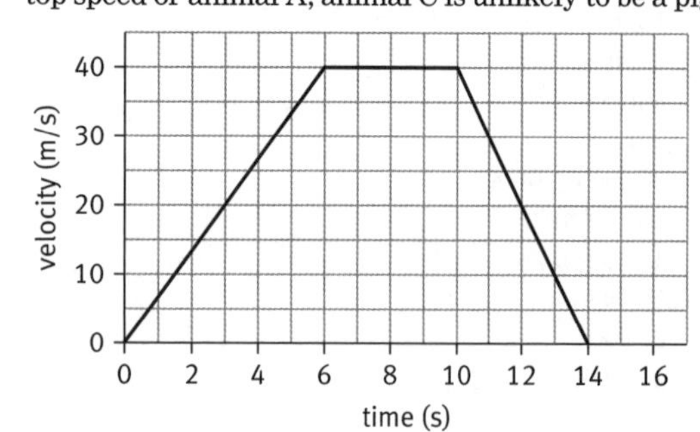

b 6.7 m/s

c The velocity changes from 30 m/s in one direction to 30 m/s in the opposite direction.
d i 100 m
ii The motorbike is travelling at a steady speed of 13.3 m/s

P5 Workout

1 Clockwise from top: volts, the same as, voltage, push, less
2 a $A_1 = A_2 = 100$ mA
b i Resistor on right
ii Greatest; ... more work is done by charge flowing through a large resistance than through a small one
c i 0.6 V
ii ...the p.d.s across the components add up to the p.d. across the battery
3 a B
b A
c B
4 1 resistance, 2 current, 3 ohm, 4 power, 5 V, 6 parallel, 7 voltmeter, 8 Ω, 9 generator, 10 induction, 11 R, 12 direct, 14 A, 15 commutator, 16 DC

P5 Quickfire

1 Ammeter (A); voltmeter (V); cell;
power supply (battery) or (DC)
or (AC); lamp; switch;
LDR; fixed resistor;
variable resistor; thermistor

2 All conductors...contain charges that are free to move; insulators...do not conduct electricity; insulators... do not contain charges that are free to move; metal conductors... contain charges that are free to move; metal conductors... contain electrons that are free to move; in a complete circuit... charges are not used up; in a complete circuit...the battery makes free charges flow in a continuous loop.
3 Hotter, more, smaller
4 a X
b Y
c Y
d B
e B
f X
5 Coil, current, out, pole
6

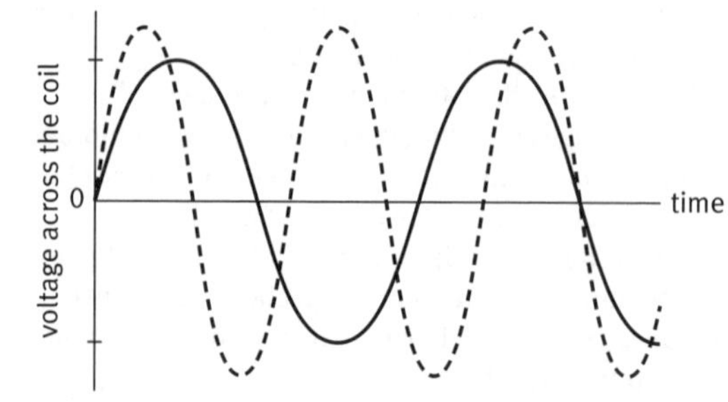

7 a 920 W
b 207 W
8 11.5 V
9 a Resistors get hotter when a current flows through them because there are collisions between the moving charges and the stationary ions in the wire.
b The potential difference is largest across the component with the greatest resistance because more work is done by the charge moving through a large resistance than through a small one.

c Mains electricity is supplied as a.c. because it is easier to generate than d.c. and simpler to distribute over long distances.

P5 GCSE-style questions

1 a i Voltmeter connected in parallel across the heater.
ii 1.2 Ω
b The wire gets hotter because moving electrons bump into stationary ions in the wire.

2 5/6 marks: answer clearly explains why the coil rotates continuously **and** the explanation is logical and coherent. All information in the answer is relevant, clear, organised, and presented in a structured and coherent format. Specialist terms are used appropriately. There are few, if any, errors in grammar, punctuation, and spelling.
3/4 marks: answer explains why the coil rotates continuously **but** the answer lacks detail **or** the explanation lacks logic and coherence.
Most of the information is relevant and presented in a structured and coherent format. Specialist terms are usually used correctly. There are occasional errors in grammar, punctuation, and spelling.
1/2 marks: answer briefly explains why the coil rotates continuously and the answer lacks detail and the explanation lacks logic and coherence.
There may be limited use of specialist terms. Errors of grammar, punctuation, and spelling prevent communication of the science. Answer includes 1 or 2 points of those listed below.
0 marks: insufficient or irrelevant science. Answer not worthy of credit.
Relevant points include:
- There are forces on the sides of the coil, because the currents in these are at right angles to the magnetic field lines.
- One of these forces is up, and the other is down, because the currents in the two sides of the coil are in opposite directions.
- These forces make the coil turn.
- The commutator swaps the current direction every time the coil is vertical.
- This reverses the forces, and makes the coil rotate continuously.

3 a Three from: increase the number of coils, increase the strength of the magnet, put an iron core inside the coil, unwind/spin the magnet more rapidly
b i The two coils of wire are wound around an iron core. A changing current in one coil causes a changing magnetic field in the iron core, which in turn induces a changing potential difference across the other transformer coil. The coil on the left has more turns, so the current in this coil is greater.
ii 120

4 a i 230 V
ii 230 V
b 3.3 amps
c Stays the same
d i 14.4 V
ii Fridge: current is smallest through this appliance, whilst the voltage across all the appliances is the same.

5 a Resistor J 3 Ω; K 30 Ω; L 60 Ω; M 2 Ω
b i Resistance $= \frac{1}{(0.2 \div 12)} = 60\ \Omega$
ii Resistor L
c The resistance calculated from the graph is 2 Ω. This increases confidence that the resistance calculated in part (a) is correct, since the value is the same.

6 a 0.025 A
b The potential difference across the three components add up to the potential difference across the battery because the work done on each unit of charge by the battery must equal the work done by it on the circuit components.
c The buzzer, because the potential difference is largest across the component with the greatest resistance.
d The potential difference across the buzzer decreases.
e i The resistor
ii In Mary's circuit, the p.d. across the resistor was smaller than the p.d.s over the other components, showing that the resistor had the smallest resistance. In the parallel circuit the p.d. across all the components is the same. The largest current flows through the component with the smallest resistance.

P6 Workout

1 Alpha: slow down, least, most, most
Beta: smaller, less, further, less
Gamma: least, least

2
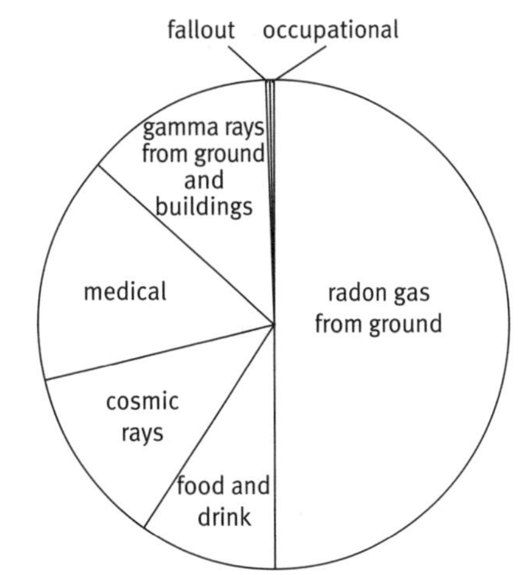

3 From top: nucleus, made up of protons and neutrons; electrons.

4
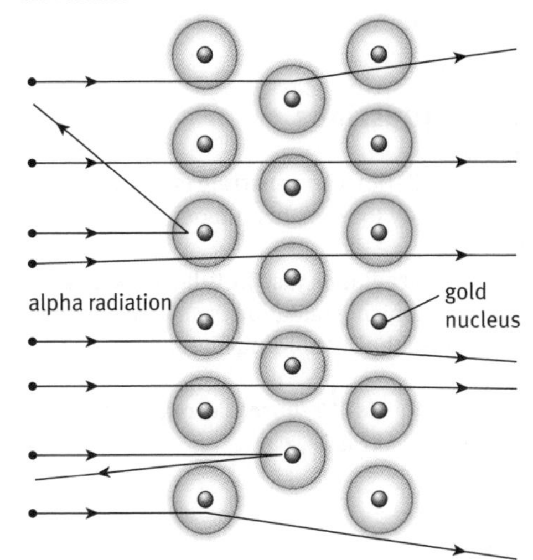

5 a A and B; D and E
b F
c E
d B and C

P6 Quickfire

1 a, b, c, d
2 True statements: **a**, **c**, **e**
Corrected versions of false statements:
- **b** Atoms of carbon-14 are radioactive. If a carbon-14 atom joins to oxygen atoms to make carbon dioxide, the carbon dioxide will be radioactive.
- **d** Radiation dose is measured in sieverts.
- **f** Hydrogen nuclei can join together to make helium nuclei. The process is called nuclear fusion.
- **g** The energy released in a nuclear reaction is much greater than the energy released in a chemical reaction involving a similar mass of material.

3 Irradiation – being exposed to radiation from a source outside your body; contamination – having a radioactive substance inside your body; radiation dose – a measure of the possible harm to your body caused by radiation.
4 Hospital radiographers, nuclear power station workers
5 **a** If the half-life was much longer that this, the krypton would continue to emit significant amounts of ionising radiation after it has done its job.
b 26 s
c Wear protective clothing, monitor radiation exposure closely.
6 Neutrons, strong, electrostatic, unstable, decays, radiation
7 **a** 9 minutes
b 8 days
c 207 years
8 **a** 92 protons and 147 neutrons
b 92 protons and 133 neutrons
c 92 protons and 125 neutrons
9 73×10^{-6} kg
10 **a** Fuel rod: contains uranium-235 atoms, which split into two smaller parts when their nuclei absorb neutrons. The process releases energy.
b Control rod: absorbs neutrons. Moved into or out of the reactor to control the reaction rate.
c Coolant: is heated up by the fuel rods, and used to heat water to make steam to turn turbines.
11 **a** ${}^{239}_{92}\text{U} \longrightarrow {}^{239}_{93}\text{Np} + {}^{0}_{-1}\beta$
b ${}^{14}_{6}\text{C} \longrightarrow {}^{14}_{7}\text{N} + {}^{0}_{-1}\beta$
c ${}^{235}_{92}\text{U} \longrightarrow {}^{231}_{90}\text{Th} + {}^{4}_{2}\alpha$
d ${}^{209}_{83}\text{Po} \longrightarrow {}^{205}_{81}\text{Tl} + {}^{4}_{2}\alpha$
e ${}^{219}_{86}\text{Rn} \longrightarrow {}^{215}_{84}\text{Bi} + {}^{4}_{2}\alpha$

P6 GCSE-style questions

1 **a** **i** Radioactive
ii Gamma radiation can penetrate deep inside Arthur's body to reach his tumour.
b **i** Kill them
ii The gamma radiation also kills healthy cells.
c **i** To prevent ionising radiation from leaving the room.
ii To minimise the dose of ionising radiation she/he receives.
2 **a** The activity of the Cs-137 source decreases over time; the half-life of Cs-137 is 30 years.
b 0.625 g
c Unstable, stable
3 **a** D, A, C, G
b Rate, boron, neutrons
c Low level – pack it in drums; medium level – mix it with concrete; high level – store in a pool of water.
4 **a** **i** The bar charts show that, for all states, the activity of Sr-90 was lower in teeth from people born between 1986 and 1989 than in the teeth from people born between 1982 and 1985. From 1986–9 onwards, the activity of Sr-90 increased.
ii No teeth were collected from children born between these years in Pennsylvania.
b The activity of Sr-90 increased from 1986–9 onwards, as did the amount of electricity generated in nuclear power stations. So the data support the conclusion.
c The finding makes the conclusion more likely to be correct. It provides further evidence that increased levels of Sr-90 are caused by the generation of electricity from nuclear sources.
5 **a** **i** It decays by emitting alpha particles, which are stopped by the skin. Alpha particles only cause harm if they are emitted by a radioactive source inside the body. (Also, plutonium is highly toxic.)
ii Two protons and two neutrons
b ${}^{238}_{94}\text{Pu} \longrightarrow {}^{234}_{92}\text{U} + {}^{4}_{2}\alpha$
c About 90 years
d 94 protons and 145 neutrons
6 5/6 marks: answer describes several ways in which ionising radiation is used in hospitals **and** includes a detailed, logical, and coherent explanation for each use.
All information in the answer is relevant, clear, organised, and presented in a structured and coherent format. Specialist terms are used appropriately. There are few, if any, errors in grammar, punctuation, and spelling.
3/4 marks: answer describes one or two ways in which ionising radiation is used in hospitals **and** includes an explanation for each use **but** the explanations lacks detail, logic, and coherence **or** includes an explanation for one use **and** the explanation is detailed, coherent, and logical. Most of the information is relevant and presented in a structured and coherent format. Specialist terms are usually used correctly. There are occasional errors in grammar, punctuation, and spelling.
1/2 marks: answer describes one way in which ionising radiation is used in hospitals **and** gives an explanation for this use **or** the answer describes one method in which ionising radiation is used in hospitals **but** does not include an explanation for this use.
There may be limited use of specialist terms. Errors of grammar, punctuation, and spelling prevent communication of the science. Answer includes 1 or 2 points of those listed below.
0 marks: insufficient or irrelevant science. Answer not worthy of credit.
Relevant points include:
- Ionising radiation is used to treat cancer by radiotherapy.
- Ionising radiation is directed at the tumour. It damages cancerous cells and makes them stop growing.
- Ionising radiation is used to sterilise surgical instruments.
- The ionising radiation kills bacteria on the surgical instruments.
- Ionising radiation is used as a tracer.
- For example, krypton-81m gas shows doctors how gases move in diseased lungs.

P7.1 Workout

1 Position 5 – full moon – light circle. Position 1 – new moon – dark circle. Position 3 – first quarter – circle with left half shaded. Position 7 – last quarter – circle with right half shaded.
2 **a** See diagram in P7.1 Factbank.
b Months 3–5

P7.1 Quickfire

1 a Planets
  b East to west
  c 24
  d Solar
2 Time for Earth to rotate once about its axis – 23 hours and 56 minutes; time for Earth to complete one orbit of the Sun – 365¼ days; time for the Moon to move across the sky once – 24 hours and 49 minutes; time for the Moon to orbit the Earth – about 28 days; time for the Sun to next reappear in the same place in the sky as it is now – 24 hours
3 True statements: b, f, i. Corrected versions of false statements:
  a During a solar eclipse the Moon comes between the Earth and the Sun.
  c During the night, stars in the northern hemisphere appear to move in circles about the Pole Star.
  d The Moon can be seen during the night and the day, depending on its rise and set times.
  e In a lunar eclipse the Earth's shadow blocks sunlight from reaching the Moon.
  g Two angles are needed to pinpoint the position of a star at any particular time.
  h Eclipses of the Moon are more frequent than eclipses of the Sun.
4 From smallest: Earth's Moon, planet, Sun, solar system, galaxy, Universe
5 a i Moon rise times get later each day.
    ii The Earth rotates on its axis once every 23 hours and 56 minutes. The Moon orbits the Earth from west to east in about 28 days.
  b 21 May
  c 7 May, since it is 14 days before the new Moon, and it takes about 28 days for the Moon to orbit the Earth.
  d Thin crescent on right side of circle.
  e The Moon has already set – it is over the horizon.
  f i On a given day, the Moon rises earlier in Ulaanbaatar than in London.
    ii Because the Earth spins on its axis. Ulaanbaatar arrives at a point where the Moon can be seen before London arrives at this point.
  g It might be cloudy or she might be looking in the wrong direction.

P7.1 GCSE-style questions

1 a Constellation
  b Angle of declination and angle of right ascension
  c The Earth has rotated on its axis.
  d It is the Pole Star, directly above the axis of rotation / the North Pole.
  e After 6 months, the Earth has moved halfway around its orbit. Orion is now in the direction of the Sun and can't be seen due to the sunlight.
  f The Moon orbits the Earth.
2 a Observe the position of the object/ over several nights. / If its position changes relative to the fixed stars then it is a planet.
  b A star is a source of light and heat / is luminous. A planet orbits a star / is seen by reflected light. Both star and planet must be mentioned to get both marks.
3 5/6 marks: answer compares the two diagrams in detail **and** comes to a reasoned conclusion about which is a better representation. All information in the answer is relevant, clear, organised, and presented in a structured and coherent format. Specialist terms are used appropriately. Few, if any, errors in grammar, punctuation, and spelling.
3/4 marks: answer compares some aspects of the two diagrams **and** comes to a reasoned conclusion about which is better. Most of the information is relevant and presented in a structured and coherent format. Specialist terms are usually used correctly. There are occasional errors in grammar, punctuation, and spelling.
1/2 marks: answer makes one or two comparisons **but** does not state which diagram they believe to be better. There may be limited use of specialist terms. Errors of grammar, punctuation, and spelling prevent communication of the science. Answer includes 1 or 2 points of those listed below.
0 marks: insufficient or irrelevant science. Answer not worthy of credit.
Relevant points include:
- Diagram 1 shows the Sun orbiting Earth, which is incorrect. Diagram 2 does not show an orbit of the Sun or Earth.
- Both diagrams show the Moon orbiting the Earth, which is correct.
- Both diagrams show that a solar eclipse is caused when the Moon is between the Earth and the Sun.
- Diagram 2 shows rays of light being emitted from the Sun, and how these form the edge of a shadow of the Moon on the Earth.
- Diagram 1 does not show rays of light being emitted from the Sun, but it does show areas of shadow on the Earth and Moon.

P7.2 Workout

1 Correct diagrams: b, d, f, h
2 From left to right: 3, 1, 2
3 a Diagram showing image marked 5 cm from lens, 3 cm below principal axis
  b Inverted, magnified

P7.2 Quickfire

1 Dioptre – the unit for measuring the power of a lens; convex lens – a lens that is thicker in the centre than the edges, causing light rays to converge; spectrum – the continuous band of colours seen when light passes through a prism; magnification – how much bigger an image is than the object; focal length – the distance between the focus and the centre of a lens; objective lens – in a telescope, the lens that is nearer the object
2 a Reflection
  b Refraction
  c Diffraction
3 True statements: a, e, f. Corrected versions of false statements:
  b When light waves travel from air to glass, they slow down.
  c The frequency of a wave cannot change once it has been made.
  d The wavelength of a wave changes when it passes from one medium to another.
4 Diffraction is least in diagram B because the gap is bigger than those in the other diagrams and the waves have the smallest wavelength.
5 a O
  b E
  c B
  d O
  e E
6 a 2, 5, 10, 2.5 (Hint: don't forget to change the cm to metres.)
  b Guy's
  c Nikhita's and Guy's
  d Nikhita's

7 Concave, reflecting, lenses, easier

8

Focal length of objective lens (cm)	Focal length of eyepiece lens (cm)	Magnification
20	5	4
30	4	7.5
25	3	8.3

P7.2 GCSE-style questions

1 a A marked as objective lens. Horizontal line marked as principal axis. Distance from centre of lens to place where rays cross marked as focal length of A.
b The stars are a very long way away so the angle they make at the telescope is too small to be measured.
c 1/0.80 = 1.25 dioptres
d 80/5 = 16
e Focal length of objective lens = 20 × 2 cm = 40 cm. Alex: distance between lenses = 40 cm + 2 cm = 42 cm. Rebekah: distance between lenses = 80 cm + 5 cm = 85 cm. So Rebekah's telescope is longer.

2 a i A
ii The width of the gap is the same as the wavelength.
b i Arecibo, because it has the greatest aperture.
ii Parkes has a bigger aperture, so the waves were diffracted least, causing sharper signals.

P7.3 Workout

1 Graph labels: vertical – speed of recession; horizontal – distance of galaxy from Earth. Period, variable, distance, light, away, Universe, redshift, galaxy, speed, graph, Hubble, uncertain, measure, accurately

2 1 Milky Way, 2 nebula, 3 Shapley, 4 Cepheid variable, 5 light-year, 6 temperature, 7 redshift. Leavitt; a method of measuring the distance to galaxies.

P7.3 Quickfire

1 Parallax – the way stars seem to move over time relative to more distant ones; parsec – the distance to a star with a parallax angle of 1 second of arc; light-year – the distance that light travels in 1 year; nebula – name once given to any fuzzy object seen in the night sky; galaxy – a group of thousands of millions of stars; Cepheid variable – a star whose brightness varies periodically

2 a A
b B
c 5.88 pc

3 a Parallax angle or observed intensity
b Luminosity
c Distance from Earth
d Luminosity
e Luminosity and distance from Earth

4 Shapley: I think they are clouds of gas within the Milky Way. Hubble: I have used a Cepheid variable star to measure the distance to one nebula. My measurements show that it is much further away than any other stars in our galaxy. So Curtis must be right.

5 a From top: 4.00, 3.5, 2.70
b The parallax angles are too small to be measured.
c 4.3 light-years

6 True statements: b, e, h. Corrected versions of false statements:
a The parallax angel of a star is half the angle apparently moved against a background of very distant stars in 6 months.
c A megaparsec is one million parsecs.
d Typical interstellar distances are measured in parsecs.
f The greater the period of a Cepheid variable, the greater its luminosity *or* the smaller the period of a Cepheid variable, the smaller its luminosity.
g The Sun is one of millions of stars in the Milky Way galaxy.
i There is no relationship between the luminosity of a star and its distance from Earth.
j Scientists believe the Universe began with a big bang about 14 thousand million years ago.

7 A, D, C, B

8 a 6000 km/s
b 1×10^{21} km
c $4.6 \times 10^{-18}\,s^{-1}$

P7.3 GCSE-style questions

1 a Cepheid variable
b Answer in range 2.0–2.1 days
c Answer in range 2050–2100 units
d 5/6 marks: answer clearly and correctly explains how the luminosity **and** distance data in table A explain the data in table B **and** the explanation is detailed. All information in the answer is relevant, clear, organised, and presented in a structured and coherent format. Specialist terms are used appropriately. Few, if any, errors in grammar, punctuation, and spelling.
3/4 marks: answer correctly explains how the luminosity **and** distance data in table A explain the data in table B **but** the explanation lacks detail. **Or** answer correctly explains how the luminosity **or** distance data help to explain the data in table B **and** the explanation is detailed. Most of the information is relevant and presented in a structured and coherent format. Specialist terms are usually used correctly. There are occasional errors in grammar, punctuation, and spelling.
1/2 marks: answer explains how the luminosity **and** distance data in table A explain the data in table B **but** the explanation lacks detail and clarity. There may be limited use of specialist terms. Errors of grammar, punctuation, and spelling prevent communication of the science. Answer includes 1 or 2 points of those listed below.
0 marks: insufficient or irrelevant science. Answer not worthy of credit.
Relevant points include:
- The observed intensity depends on the luminosity of a star and its distance from Earth.
- For stars at the same distance from Earth, as luminosity increases, so does observed intensity.
- For stars of the same luminosity, as the distance from Earth increases, the observed intensity decreases.
- If all the stars in table B were of the same luminosity, the order of observed intensity would be Epsilon Cassiopeiae, Alpha Cassiopeiae, Beta Cassiopeiae.
- But both factors must be taken into account to explain observed intensity. For example, the fact that Alpha is closer to Earth than Epsilon is is more significant than Epsilon's greater brightness in determining their observed intensities.
- The fact that Beta is closer to Earth than Epsilon is is more significant than Epsilon's greater brightness in determining their observed intensities.

2 a Parallax is the apparent shift of an object against a more distant background as the position of the observer changes.
b Sirius A is closest to Earth because it has the biggest parallax angle.

c 1/0.286 = 3.50 parsecs

d i Alpha Cygni is further from Earth than 61 Cygni is. This shows that the distances from Earth of the stars in a constellation can be very different from each other.

ii Data on the distances from Earth of other stars in the constellation

3 a A galaxy is a collection of thousands of millions of stars held together by gravity.

b Milky Way

c In another galaxy, because distances between galaxies are measured in megaparsecs.

d $2.3 \times 10^{-18}\ s^{-1}$

e It is hard to get an accurate measurement for the distance to a very distant galaxy.

P7.4 Workout

1 Gravity pulls gases together to form a protostar.
Main sequence star fusing hydrogen into helium.
Red giant fusing helium into larger nuclei.
White dwarf gradually cools.
Red supergiant fusing helium and other nuclei to make much larger nuclei up to iron.
Supernova
Neutron star
Black hole

2 Correct diagram is C.

3 Correct diagrams are B and C.

P7.4 Quickfire

1 a Top line – hottest star, bottom line – coolest star

b Colours from top: white, yellow, red

2

Temperature (°C)	Temperature (K)
0	273
200	473
−200	73
−273	0
−266	7
27	300

3 a 300 cm^3

b 5 dm^3

4 2×10^5 Pa

5 1333 Pa

6 True statements: a, d

7 a In space, gravity compresses a cloud of hydrogen and helium gas.

b Correct

c Correct

d The volume of the cloud has decreased.

e As the gas particles fall towards each other they move more and more quickly, so the temperature and pressure increase.

8 a $^{12}_{6}C + ^{1}_{1}H \rightarrow ^{13}_{7}N$

b $^{13}_{7}N \rightarrow ^{13}_{6}C + ^{0}_{+1}e$

c $^{13}_{6}C + ^{1}_{1}H \rightarrow ^{14}_{7}N$

d $^{15}_{7}N + ^{1}_{1}H \rightarrow ^{12}_{6}C + ^{4}_{2}He$

9 Box 1: a, f. Box 2: b, d, g. Box 3: c, e

10 See Hertzsprung–Russel diagram in P7.4 Factbank.

P7.4 GCSE-style questions

1 5/6 marks: answer clearly refers to each piece of evidence and draws a sensible overall conclusion based on the evidence **and** points out uncertainties in the conclusion **and** identifies any evidence not used. All information in the answer is relevant, clear, organised, and presented in a structured and coherent format. Specialist terms are used appropriately. Few, if any, errors in grammar, punctuation, and spelling.
3/4 marks: answer refers to two or three pieces of evidence and draws a sensible overall conclusion based on the evidence **and** points out uncertainties in the conclusion **or** identifies any evidence not used. Most of the information is relevant and presented in a structured and coherent format. Specialist terms are usually used correctly. There are occasional errors in grammar, punctuation, and spelling.
1/2 marks: answer refers to one piece of evidence and suggests what this piece of evidence shows. There may be limited use of specialist terms. Errors of grammar, punctuation, and spelling prevent communication of the science. Answer includes 1 or 2 points of those listed below.
0 marks: insufficient or irrelevant science. Answer not worthy of credit.
Relevant points include:

- Evidence A suggests that the star is a main sequence star.
- Evidence B suggests that the star is a main sequence star, since it is mostly hydrogen and helium.
- However, the star also includes other elements – these suggest that the star might not be a main sequence star or...
- ...that the star was created from dust expelled by a supernova.
- Evidence C supports evidence A and B by showing that the star is a main sequence star.
- The lines on the absorption spectrum for star X are also present in the spectrum for hydrogen...
- ...This suggests that star X contains hydrogen, and is a main sequence star.
- There are lines on the absorption spectrum of star X that are not on the hydrogen spectrum...
- ... From the evidence given, it is not possible to identify the element they are from.

2 a Positron

b 2

c He

d i Reaction 1: mass change = 0.00101 u
Reaction 2: mass change = 0.00590 u

ii Reaction 3 releases most energy. The mass change is greatest for this reaction. Since $E = mc^2$ the reaction with the greatest change in mass will also have the greatest change in energy.

e 4.3×10^9

P7.5 Workout

1 Disadvantages of space telescopes are that they are – expensive to set up and maintain; computers are used to control telescopes because they can – enable a telescope to track a distant star while the Earth rotates, allow the telescope to be used by an astronomer not at the observatory; international cooperation in astronomy allows – the cost of new major telescopes to be shared, the pooling of scientific expertise; in deciding where to site a new observatory it is necessary to consider – the amount of light pollution, common local weather conditions, the environmental and social impact of the project.

2 a Several miles from the smoke of London

b The land did not need to be purchased.

c Light from cities / other sources on Earth which make it harder to see astronomical objects

d Cloud / mist / rain obscuring the night sky

e Above the clouds; less affected by refraction from the atmosphere

P7.5 Quickfire

1 True statements: b, c, d, f
Corrected versions of false statements:
  a Space telescopes do not necessarily have bigger lenses or mirrors than Earth-based ones.
  e Light pollution makes it harder to see stars.
2 Australia, Canary Islands, Chile, Hawaii
3 a S
  b O
  c O
  d S
  e S
  f S
  g O
  h O
  i O
  j S
4 Data, travel, track, positioned, record, process, communicate
5 a Reflector – it has a mirror.
  b Expertise and cost can be shared.
  c It is above the atmosphere; it avoids absorption and refraction effects of the atmosphere
6 a Gran Telescopio Canarias, because it has the largest aperture, so can collect the most light
  b Gran Telescopio Canarias, Southern African Large Telescope
7 Keck 1 and Subaru

P7.5 GCSE-style questions

1 a 5/6 marks: answer compares many advantages **and** disadvantages of each site **and** comes to a reasoned conclusion, based on evidence, about which site is better. All information in the answer is relevant, clear, organised and presented in a structured and coherent format. Specialist terms are used appropriately. Few, if any, errors in grammar, punctuation, and spelling.
  3/4 marks: answer compares some advantages and disadvantages of one site **or** some advantages of both sites **or** some disadvantages of both sites **and** states which site is better **but** reasons for conclusion lack detail / do not refer to advantages and disadvantages identified. Most of the information is relevant and presented in a structured and coherent format. Specialist terms are usually used correctly. There are occasional errors in grammar, punctuation, and spelling.
  1/2 marks: answer points out one or two advantages or disadvantages of one or both sites site **but** does not state which site they believe to be better. There may be limited use of specialist terms. Errors of grammar, punctuation, and spelling prevent communication of the science. Answer includes 1 or 2 points of those listed below.
  0 marks: insufficient or irrelevant science. Answer not worthy of credit.
  Relevant points include:
  - Astronomical advantage of Leh – it is at a higher altitude than Mumbai.
  - Astronomical advantage of Leh – the climate suggests that it is less cloudy than Mumbai.
  - Astronomical advantage of Leh – the town is much smaller than Mumbai, so there is likely to be less light pollution.
  - Astronomical advantage of Leh – the town is smaller so its air is likely to be less polluted than that of Mumbai.
  - Disadvantage of Leh – cost of building telescope and of bringing in supplies for workers is high since it is inaccessible in winter.
  - Disadvantage of Leh – fewer potential workers compared to Mumbai, which has many computer experts and a higher proportion of literate people.
  - Disadvantage of Leh – might cause visual pollution in a beautiful area.
  b Number of cloudy nights in each location, actual levels of air pollution
2 a Four from: supernovae, neutron stars, black holes, main sequence stars, remains of supernovae
  b AGILE and RadioAstron
  c Hubble and Herschel
  d Costs and expertise are shared.

Ideas about science 1 Workout

1 The temperature in the main part of the room is different from the temperature behind the curtain; the thermometer may be inaccurate; he may read the thermometer incorrectly.
2 A – 4, B – 2, C – 3, D – 1, E – 6, F – 7, G – 5

Ideas about science 1 GCSE-style questions

1 a 25.0 and 32.1
  b i 32.1
    ii If possible, she should check the result again.
  c i Tillie's data set has the bigger range for voltage values, because her range is 1.5–7.5 V (6.0 V difference between the highest and lowest values) and Charlie's range is 3.0–5.0 V (2.0 V difference between highest and lowest values).
    ii True value = mean = 26 Ω
  d The mean of Tillie's values lies within the range of Charlie's values, and the mean of Charlie's values lies within the range of Tillie's values. So the true value is likely not to have changed.

Ideas about science 2 Workout

1 1 causal, 2 outcome, 3 factor, 4 fair, 5 mechanism, 6 flawed, 7 chance, 8 matched, 9 increases, 10 random, 11 control

Ideas about science 2 GCSE-style questions

1 a Volume of gas
  b i Temperature of gas, pressure of gas, mass of gas
    ii Pressure of gas, mass of gas
    iii To make sure the test is fair
  c

Temperature (°C)	Height of column (mm)	Volume of air (mm^3)
0	0	80
15	18	94
20	24	99
25	32	106
31	47	118
36	58	126

    ii Temperature scale on x-axis: 0–40 °C. Volume scale on y-axis: 80–130 mm^3. Points plotted correctly. Line of best fit sloping from bottom left to top right.
    iii There is a linear relationship between volume and temperature.
    iv As temperature increases, the particles have more energy and move around more quickly. They spread out, which increases the volume.
2 a As temperature increases, luminosity increases.
  b i Outcome – luminosity; input factor – surface area
    ii For a given temperature of star, the luminosity of a red giant is greater than that of a main sequence star because the surface area of a red giant is greater. Therefore, for the temperature range of red giants, red giants appear above main sequence stars on the Hertzsprung–Russell diagram.

Ideas about science 3 Workout

1 Statements that are true: c, d, f, h

Corrected versions of statements that are false:

- a Scientific explanations have to be thought up creatively from data.
- b An explanation may be incorrect, even if all the data agree with it.
- e An explanation may explain a range of phenomena that scientists didn't know were linked.
- g If an observation agrees with a prediction that is based on an explanation, it increases confidence that the explanation is correct.
- i If an observation does not agree with a prediction that is based on an explanation, then the explanation or the observation may be wrong. Confidence in the explanation is reduced.
- j If an observation does not agree with a prediction that is based on an explanation, then the explanation or the observation may be wrong. Confidence in the explanation is reduced.

Ideas about science 3 GCSE-style questions

1 If they find iridium, this will increase confidence in the explanation.

If they do not find iridium, the prediction may still be correct.

2 a i 1 C and H; 2 A; 3 E and F; 4 D; 5 G and I; 6 B

ii This does not mean the explanation is correct, since there could be another explanation that also accounts for all the data.

b It is impossible to predict when the next Earth movement will be.

3 a 1 D, 2 D, 3 D, 4E

b They could not see the nucleus; any other sensible suggestion.

c The fact that the prediction is correct increases confidence in the explanation; if the prediction were wrong, we would be less confident in the explanation.

d Energy being emitted when uranium atoms decay; *or* the total mass of the products of the decay reaction being less than the mass of the original uranium.

4 a 1 A, B, C, D; 2 E; 3 F; 4 G

b Galileo made the groove as smooth as possible to reduce friction.

c He repeated the measurements many times to check their repeatability or to obtain values that are as close to the true values as possible.

d i He could repeat the test described, but with slopes of different steepness.

ii The ball would have speeded up more on a steeper slope.

Ideas about science 4 Workout

1 1 E, 2 C, 3 B, 4 D, 5 G, 6 A, 7 F, 8 H

2 Examples of suitable sentences include:

- Scientists are usually sceptical about unexpected findings until they have been replicated by themselves or reproduced by others.
- Scientists report their claims to other scientists through scientific conferences and journals.
- Peer review is the process of critical evaluation of scientific claims by other scientists who are experts in the same field.
- There is less confidence in a new scientific claim that has not yet been evaluated by the scientific community than there is in well-established claims.

Ideas about science 4 GCSE-style questions

1 a So that other scientists can try to reproduce the findings in future; so that other scientists can ask questions about the findings

b i The shapes of the continents seemed to fit together like a jigsaw; the rock types of eastern South America are the same as those of western Africa; the fossils of eastern South America are similar to those of western Africa.

ii The idea of continents moving was outside their experience; they could not imagine how the continents could move; Wegener was not regarded as a member of the community of geologists.

iii Data collected since 1912 support the explanation; data collected since 1912 agree with predictions based on the explanation.

Ideas about science 5 Workout

1 1 harm; 2 chance; 3 advance; 4 ionising; 5 consequences; 6 assess; 7 controversial; 8 statistically; 9 benefit; 10 perceive; 11 acceptable

2 Sentences might include the following:

- People's perceptions of the size of a particular risk may be different from the statistically estimated risk.
- The perceived risk of an unfamiliar activity is often greater than the perceived risk of a more familiar activity.
- Governments have to assess what level of risk is acceptable in a particular situation. This decision may be controversial.

Ideas about science 5 GCSE-style questions

1 a Zion and Junaid

b Zion, Junaid and Sarah

c Catherine

2 a Government energy minister – nuclear power stations provide electricity for many people; Pete – the extra cancer risk is significant, but there are few jobs in the area and the pay is good; Jake – the calculated risk of his getting cancer is small, but if it happened the consequences would be terrible.

b The nuclear reactors are shielded from the workers.

c The risk of harm to those living closer to the power station is greater.

d Workers at the French power stations, and those living near them, would be at increased risk of harm. There would be less potential harm to people in the UK, since there would be no nuclear power stations. Nuclear workers in the UK might lose their jobs.

3 a Tom is correct – the data show that the typical annual radiation dose of a nuclear power station worker is less than that of a typical pilot.

b Barbara is more familiar with flying, so her perception of the size of the risk is less than the statistically calculated risk. Barbara is afraid of the invisible radiation from power stations, but perhaps is less aware that flying also exposes people to ionising radiation.

4 a Two points from: observations used to find out more about Cepheid variable stars, black holes, and the expansion of the Universe; images sent to Earth, including collision of comet with Jupiter

b i For example: the Space Shuttle they travel on may break up on re-entry to Earth's atmosphere, and kill them.

ii Astronomers / scientists (other answers are possible)

iii Families of astronauts (other answers are possible)

c 5/6 marks: answer discusses the benefits and risks of each option in detail **and** states, with detailed reasons, which option they believe to be best. All information in the answer is relevant, clear, organised, and presented in a structured and coherent format. Specialist terms are used appropriately. Few, if any, errors in grammar, punctuation, and spelling.

3/4 marks: answer discusses the benefits and risks of one or two options **or** the benefits of some options and the risks of other options **and** states with brief reasons which option they believe to be best. Most of the information is relevant and presented in a structured and coherent format. Specialist terms are usually used correctly. There are occasional errors in grammar, punctuation, and spelling.

1/2 marks: answer discusses the benefits **or** risks of one option or the benefits of one option **and** the risks of one option **or** states with brief reasons which option they believe to be best. There may be limited use of specialist terms. Errors of grammar, punctuation, and spelling prevent communication of the science. Answer includes 1 or 2 points of those listed below.

0 marks: insufficient or irrelevant science. Answer not worthy of credit.

Relevant points include:

- Option 1 benefit – tried and tested method of servicing the HST to a high standard.
- Option 1 benefit – humans can respond to changing situations and make decisions about how best to service the HST in the light of their observations when they arrive.
- Option 1 risk – risk to life and health of astronauts during the whole mission.
- Option 1 benefit – risk above mitigated by possibility of travelling on to International Space Station if an in-flight problem develops.
- Option 2 benefit – less risk to human life and health.
- Option 2 risk – time needed to develop robots for this purpose, during which the HST may develop further faults.
- Option 3 benefit – no risk to life and health of humans, and low cost.
- Option 3 risk – the HST might develop faults or stop working altogether, so preventing its further contributions to an understanding of astronomy.

5 a Vision improving

b Risk of cornea damage / needing a cornea transplant; risk that it would be very difficult to get used to the device.

Ideas about science 6 Workout

A person or people who identify...	Name or names
...a scientific reason for building the telescope in Addis Ababa.	Emebet, Tamrat
...a scientific reason against building the telescope in Addis Ababa.	Mesfin
...a social impact of building the telescope in Addis Ababa.	Ayalnesh
...an ethical issue.	Miriam
...an issue linked to sustainability.	Workeneh
...a financial issue.	Workeneh
...a claim that could be investigated scientifically.	Emebet, Tamrat, Mesfin

Ideas about science 6 GCSE-style questions

1 a i Ian, Kris

ii Two from: Grace, Harry, Ian

iii Jasmine

b Making sure used fuel rods are disposed of safely; regularly checking storage sites where low-level and intermediate radioactive waste are stored.

2 a The energy in the waste is not simply released to the atmosphere as low-grade heat, but is used to provide useful energy. This may reduce the demand for fossil fuels.

b Costs of transporting the waste to power stations may be high; sorting waste that can be used in this way may be expensive or technically difficult.

3 5/6 marks: answer clearly identifies four or more advantages **and** disadvantages of selling the rainforest to the palm oil company **and** comes to a judgement about which is the better choice **and** gives clear reasons for the judgement, linked to the advantages and disadvantages given. All information in the answer is relevant, clear, organised, and presented in a structured and coherent format. Specialist terms are used appropriately. Few, if any, errors in grammar, punctuation, and spelling.

3/4 marks: answer identifies two or three advantages **and** disadvantages of selling the rainforest to the palm oil company **and** gives to a judgement about which is the better choice **but** does not give reasons for the judgement. Most of the information is relevant and presented in a structured and coherent format. Specialist terms are usually used correctly. There are occasional errors in grammar, punctuation, and spelling.

1/2 marks: answer identifies one or two advantages **or** disadvantages of selling the rainforest to the palm oil company **or** gives a judgement about which is the better choice **but** does not give reasons for the judgement. There may be limited use of specialist terms. Errors of grammar, punctuation, and spelling prevent communication of the science. Answer includes 1 or 2 points of those listed below.

0 marks: insufficient or irrelevant science. Answer not worthy of credit.

Relevant points include:

- Benefit – will gain financially
- Cost – forest will not be available for future generations to live in
- Cost – amount of carbon dioxide in the atmosphere will increase
- Cost – loss of biodiversity
- Benefit – burning palm oil to produce electricity will reduce the demand for fossil fuels
- Reasoned decision referring to costs and benefits.

Index

Index